THE COMPLETE ART OF FIREWORK-MAKING

Other Books by

AFN - *AMERICAN FIREWORKS NEWS*

BIBLIOGRAPHY OF FIREWORK BOOKS (Philip)

BLACK POWDER MANUFACTURE METHODS & TECHNIQUES (von Maltitz)

CHEMISTRY OF PYROTECHNICS

DICTIONARY & MANUAL OF FIREWORKS (Weingart)

FIREWORKS YEARBOOKS

FIREWORKS, History & Celebration (Plimpton)

GIRANDOLA BOOK

GLITTER, The Chemistry & Techniques (Oglesby)

INTERNATIONAL FIREWORKS TRADE DIRECTORY

NORTH AMERICAN FIREWORKS TRADE DIRECTORY

PRINCIPLES OF PYROTECHNICS (Shidlovskiy)

ROUND STARS & SHELLS (Bleser)

THE BEST OF AFN II

THE BEST OF AFN III

THE PYROTECHNIST'S TREASURY

THE COMPLETE ART OF FIREWORK-MAKING

BY

THOMAS KENTISH

A NEW IMPRESSION

WITH ILLUSTRATIONS

REPUBLISHED WITH SOME TEXT REVISONS IN 1993 BY AMERICAN FIREWORKS NEWS

The Pyrotechnist's Treasury
The Complete Art of Firework-Making
A New Impression - 1899
by Thomas Kentish

ISSN 0-929931-09-2
Rekeyed with minor text revisions
by American Fireworks News staff
HC67 - Box 30, Dingmans Ferry, PA 18328

Third printing under this copyright on the 100th anniversary of
A New Impression, March, 1999.

Warning: This text contains descriptions and depictions of fireworks. Readers are cautioned that they must form their own opinion as to the application of anything found herein.

PREFACE

TO THE 1993 EDITION

As you study this book it's hard to believe that it was written 115 years ago. To be more accurate, the first text was written in 1878, and this test - *A New Impression* - was revised from the first text and published in 1899. Regardless of the precision of the dating, the fact remains that the wonderfully complete descriptions of fireworks manipulations that the author gives are just as valid today as when he wrote them. Actually, the descriptions for some devices are more detailed than any found in the current fireworks descriptive texts. His marvelous treatment of how to assure that all Roman candle stars rise to the same height is an example of the fine details to be found here, and provided much of the persuasion we needed to decide to republish t he book.

Thomas Kentish was a working pyrotechnist in the late 19th century. We will never know why he decided to write the book; it probably had something to do with making money. Proof that he was successful can be found in the fact that his widow continued to live off the sales of the book for fifteen years after his demise. And considering the success of the book then, it is amazing that it continues to this day to be a class fireworks how-to book.

Some words of caution are needed. We know a great deal more today about the chemistry of fireworks formulations than chemist Kentish did. Although he cautioned against potassium chlorate/sulfur combinations, his formulations leaned heavily upon their use. Likewise, he advocated use of mercurous chloride and arsenic disulfide. These very toxic chemicals have been completely eliminated from our chemical inventories today. We retained those formulations in the 1993 edition merely to maintain the historic flavor of the manuscript. Readers are cautioned to view them in just that light. There are plenty of modern formulations readily obtainable from the current literature.

As we prepared this new edition, we faced the decision of going with the original book or rekeying the manuscript. The original *A New Impression* had been printed in 1899 and reproduced a few times, so the copy we had contained less than ideal printing. We decided to rekey the text and modernize certain items.

We altered the text as little as possible. Chemical names have been updated. For instance Carbonate of Strontia has become strontium carbonate. We eliminated certain tedious references, such as being able to secure wing nuts from the local ironmonger for 2d. Other than these few modernizations, the text remains loyal to Thomas Kentish's turn-of-the-century prose.

This is an amazing book that has lost little over the century. Enjoy it - the reading will transport you to the author's fireworks workshop in Victorian England. Yet you'll marvel at how little the basic techniques have changed. Fireworks then are fireworks now.

John M. Drewes
Dingmans Ferry, PA
April, 1993

PREFACE

TO THE SECOND EDITION.

The present edition contains much additional information, many fresh illustrations, and several pieces of a construction entirely new; as winged and coloured tourbillions; chain, or triplet asteroids; and the manufacture of metallic drifts. An account of the latter is placed at the end of the book, to avoid disturbing the previous arrangement; but I recommend the reader to study the chapter describing them before he commences to practice, as it will enable him to construct his own rocket apparatus without the inconvenience of delay, or the expense of purchase. For the rest, I have carefully revised the book throughout, and have supplied the minutest information in every instance in which it appeared to be defective; and, to make it as comprehensive as possible, I have given a description of aquatic fireworks, for the amusement of those who may have access to large pieces of water. The work appears to me now complete.

Charging a Case.

PREFACE

TO THE FIRST EDITION.

Until within the last few years there was scarcely any work in English, that I am aware of, on the subject of Pyrotechny, worth reading, with the exception of an Article in "Brewster's Cyclopaedia," by MacCulloch; and this, besides being accessible to only a few, having been published in 1830, made no mention of colours, which form the most beautiful part of the art.

In the year 1770 a treatise was written by a Captain Jones, which has been copied, in whole or in part, into almost every work since published. The greater portion of it is absurd and impracticable, and shows that it was written by a person who undertook to teach what he had not learnt.

The first work of any real utility that came under my notice was a series of papers by "Practicus." This was soon followed by another, varied by the new chemical nomenclature.

The subject is far from being treated exhaustively in either of these works, so that I trust the reader will find in the following pages a fund of information, both in the repertory of recipes and the methods of manipulation.

To Chertier belongs the great improvement in colours. He was, as I was informed by the late Mr. Southby, who knew him personally, and who derived much information from him while in Paris, a retired French Artillery Officer, who made colours "his study and theme." His "Nouvelles Recherches," published in 1852, nearly thirty years after his first pamphlet, is an excellent work, that leaves little to be desired in the way of colours. Gunpowder attracted but little of his attention.

Tessier, of Paris, has written, since, his "Chimie Pyrotechnique;" and a new edition of the "Lustfeuerwerkkunst," by Hutstein and Websky, was published three or four years ago at Breslau.

I have neglected neither of these sources of information, but do not know that I have been able to learn much from them with which I was not previously acquainted.

Gibbon's "Artillerist's Manual," Benton's "Ordnance and Gunnery," and Simpson's "Ordnance and Naval Gunnery," published at New York, have furnished me with a hint or two; also Holley's "Ordnance and Armor," Ward's "Naval Ordnance and Gunnery," Owen's "Practice of Modern Artillery," Scoffern's "Projectile Weapons of War," and his "New Resources of Warfare;" but as all these works are on Military Pyrotechny, they have not been available to any great extent. I mention their names, that anyone desirous of a knowledge of Rockets and Shells, as instruments of destruction, may know where to gain the information.

It is possible that, as Chemistry advances, a few new substances may be discovered: meanwhile, nearly every shade of colour may already be produced.

The extensive use of these colours, from their beauty and variety, for stars and lance-work, has very materially altered the class of fireworks, and necessitates the employment of an enormous quantity of quick-match. The preparation of this is one of the most disagreeable parts of Pyrotechny, besides demanding a great amount of manipulative skill. Most amateurs are deterred from attempting to manufacture it, and so have to content themselves with only the simplest pieces.

I have, therefore, set myself sedulously to work to devise a different plan of preparing it; and have succeeded--not till after many years, however, and numerous failures, simple as it now seems--in discovering a process of producing a splendid and perfect match, that leaves nothing to be desired; which is easy and expeditious, and does not even soil the hands. Amateurs will now find no trouble in making any pieces they may desire.

Dangerous chemicals, that might produce spontaneous combustion, have been rigidly excluded; the fullest information throughout has been furnished for preparing the mixtures with safety; the mode of projecting balloons, and the manner of constructing steelyard scales, and every kind of tool used in the art, has been added; and no pains have been spared to make the work as complete and comprehensive as possible.

In conclusion, I may say that I have had thirty-five years' experience: and there is nothing I have not tried repeatedly: the reader may, therefore, place the greatest confidence in everything recommended. He has only to imitate to succeed.

CONTENTS

Interior of Torpedo Shop

THE PYROTECHNIST'S TREASURY.

ROMAN CANDLES.

To Make a 5/8 Roman Candle.

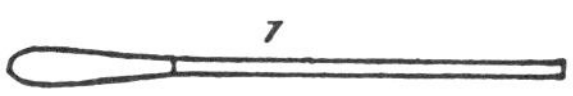

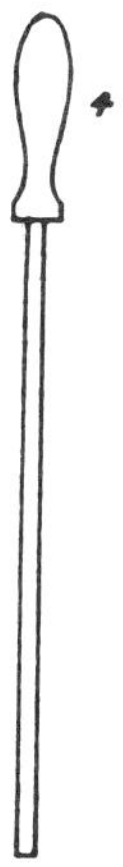

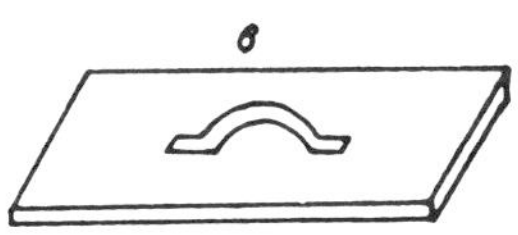

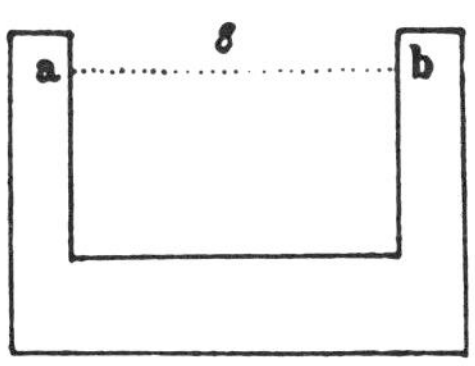

Procure a straight piece of brass tube, 5/8" external diameter and 16½" long. Saw or file off a piece, 1½" long, fig. 1. This is for the star former, or mould.

In another piece, 15" long, fix a handle, as shown in fig. 7. This is for the case former, or roller. It should be filed smooth at the end.

Take another piece of brass tube 9/16" in external diameter, and about 16½" long. In this also fix a handle, or fix it into a handle, fig. 4. Invert it, and set it upright in a flower-pot, filled with dry sand or loose mould. Melt some lead in a ladle, and pour it slowly into the tube, leaving room for the air to escape up the side, till it is full. If the lead is poured in rapidly, the confined air, expanding, jerks the metal up, and may cause serious injury. A pound or more of lead will be required. When cold, file the end smooth. This is for a rammer. A copper bolt, or brass rod, will serve instead, if preferred; iron should not be used.

Take a piece of pine, fig. 6, about 12" long, 6" wide and 3/4" thick; and, on the top, screw a handle. This is for a rolling-board. An iron door-handle would serve. A wooden one, however, about 1" thick, not cylindrical, but slightly flat and rounded at the edges, is preferable, as it gives more purchase for the hand. A good handle may be made with 8" of a cricket stump, 2 cotton reels and 2 long screws.

Cut a piece of tin, or zinc, or thin board, into the shape of fig. 8, in which the distance between the arms a and b, across the dotted line, shall be 7/8". This is for a gauge, or caliper, with which to measure the external diameter of the case. Write upon it, "7/8 space".

Procure some 60 lb., 70 lb., or 84 lb. kraft paper: the size of a sheet will be 29" by 22½". Cut a sheet into 4 equal parts, each 14½ by 11¼"; paste the 4 pieces on one side, and lay them one on another, with the pasted face upwards, putting the fourth piece with the pasted side downwards upon the pasted side of the third piece. Turn them over: take off the now top piece and lay it flat on the near edge of a table, pasted side upwards. Take the former, fig. 7, and paste the tube all over; or smear it with a little sweet oil and wipe it nearly dry. Lay it along the edge of the paper, bend the paper over with the fingers of both hands and roll it tightly up, until the external diameter of the case about fits the gauge, fig. 8. If the paper should be too long, of course a piece must be cut off; if it should not be long enough, more must be added, taking care to bind in the second piece with 3 or 4" of the first piece; for if the whole of the first piece be rolled up before beginning the second, the latter, when dry, will probably slip off and spoil the case. The case having been rolled up, take the handle of the former in the left hand, lay the case flat on the near side of the table, take the rolling-board, fig. 6, in the right hand, press the front part of it on the case, and drive it forwards 5 or 6 times, like a jack-plane, letting the handle of the former slip round in the left hand. This will tighten the case, and render it, when dry, as hard as a book-cover.

The former must always be pasted, or rubbed with an oiled rag, before rolling a case, to prevent its sticking. It should, likewise, be wiped clean with a damp sponge before being laid aside. Brass tubes keep clean a much longer time if lacquered. To lacquer them, clean them with very fine sandpaper; make them hot by the fire, till you can just bear them on the back of the hand; then, with a camel-hair pencil, wash them over with thin shellac solution. For muslin plates, or brass bands, it is advisable to purchase the lacquer ready made. The cases may be either $14\frac{1}{2}$ or $11\frac{1}{4}$" long; but $11\frac{1}{4}$" is the best, for when the cases are too long, the fuse, as it approaches the bottom, is apt, if slow, to smoke; if fierce, to set the top of the case in a flame.

After the first case has been rolled up to fit the gauge, it may be unrolled, and the paper measured. Future pieces can then be cut of the right size at once, so that the case will fit the gauge, and be of the proper thickness, without further trouble.

A large slab of slate is convenient for rolling upon; but a smoothly-planed board will answer every purpose.

When a number of cases are finished, hitch a piece of string two or three times round each of them, and hang them up to dry, in a place free from draft so that they will not warp.

Flax is sold in balls; the thick yellow is best. It is named, indifferently, flax or hemp; it is used by shoemakers. Two or three thicknesses of this, waxed, or drawn through the hand with a little paste, is very convenient for passing round the necks of small choked cases, tying cases on wheels, &c. They also sell a yellow cotton, on reels, which is equally serviceable.

To Make a Roman-Candle Star.

2

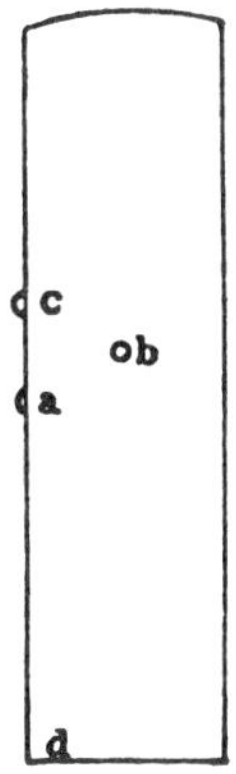

3

26

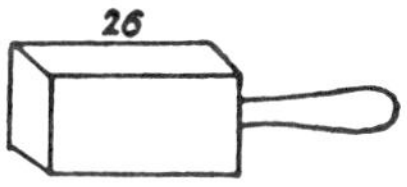

Take the former, fig. 1, which, as said before, is $1\frac{1}{2}$" long; have a cylindrical piece of turned wood, box, beech, or mahogany, fig. 2, about 2" long and of a diameter to just fit easily into fig. 1. At a point a, at the distance of 1" from the end d, with a bradawl, or very small gimlet, make a hole, and drive in a piece of brass wire to project just so much as to prevent the tube slipping over it. A piece of a brass rivet, such as used by shoemakers, fig. 209, is convenient for the purpose. The part with the head on is best, as being the smoothest; a quarter of 1" length will be sufficient, filed or cut off with the nippers. It is evident that upon inserting fig. 2 into the tube fig. 1, a vacant space of $\frac{1}{2}$" will be left at the bottom. fig. 3 is a piece of turned wood, or, better still, of turned or filed brass rod, exactly like fig. 2 without the side-pin a. Now to pump a star, insert fig. 2 in fig. 1, press the tube into damped composition till full, turn the drift round and withdraw it. Rest the tube on a flat surface, insert fig. 3, and give it two or three taps with a light hammer, or with a small mallet, like fig. 26. A convenient size for the mallet is $1\frac{1}{2}$" square, 3" long, with a turned handle. The mallet is best made of beech or mahogany. The slight malleting consolidates the star and prevents it from getting broken in charging; it will compress it to about 7/16" in height; brush the bottom of it clean, if necessary, with a sash-tool; then push it out, and set it by to dry.

It will much facilitate the delivery of the star if the interior of the tube, at its lower extremity, be filed sloping, a trifle larger, with a rat-tailed, or round file. File a notch, or mark, near the top, to distinguish it from the bottom.

Stars are best made in summer, though they may be prepared and set by, on a shelf, at any time, in a warm room. A little wooden tray, about 6" square, with a rim round it; or a pine frame, with a perforated zinc bottom, will be found convenient for holding the stars. When dry, they should be put into clean pickle-bottles, furnished with tight-fitting bungs, to protect them from the damp air. A piece of wash-leather, or flannel, passed over the bottom of the bung, gathered up round the sides and tied at the top like a choke, makes a good stopper. Shot, shaken up in bottles, with water and a little sand, soon cleans them. If they contain anything oily or acid, add soda; if anything alkaline, add oxalic acid; both previously dissolved in hot water and allowed partly to cool.

To Damp Stars

Stars containing strontium nitrate must be damped either with lac solution, or wax solution; anything containing water destroys the colour. Nitrate stars may be damped with gum water, dextrine solution, or thin starch. Most other stars with either of the solutions. Reds and greens will mix with boiled linseed oil, but they cannot then be matched, as oil renders meal-powder almost uninflammable. With all stars, not a drop more of the solution should be used than is sufficient to make the composition bind; and it is advisable not to damp more than half an ounce at a time; this is particularly the case in using the lac solution, as it dries rapidly; and if a large quantity of composition is damped and gets dry, and has to be damped over and over again, it becomes clogged with the shellac and the colour is deteriorated. If it should get dry and require a second damping, it is best to use pure alcohol only, the second time. When about to damp a very small quantity, measure the lac solution in a teaspoon.

Before mixing compositions, every article should be as fine as wheat flour and perfectly dry. Strontium nitrate, if purchased in the lump, should be set over the fire in a pipkin; it will soon begin to boil in its water of crystallization; it must be kept stirred, with a piece of wood, till the water is evaporated and a fine dry powder left. A pound of crystals will yield about 11 ounces of dry powder, which should be immediately bottled. Even then, if used in damp weather, it is best dried again and mixed with the other ingredients while warm. This second drying may be in a 6" circular frying-pan. Stars or mixtures containing strontium nitrate must never be left exposed to the atmosphere, as it is the most hygroscopic of all pyrotechnic chemicals, except sodium nitrate, which is no longer used. It is best to buy the strontium nitrate ready dried.

Articles, separately, may be reduced to powder with the pestle, in a mortar. See that it is wiped clean every time, as there is danger of ignition with chlorates and sulfurs. When the articles are to be mixed, they may be put into the mortar and stirred together with a small sash brush. A 3/8" diameter is a convenient size. The mixture must then be put into a sieve and shaken in the usual way; or it may be brushed through with the sash brush. Return it to the sieve, and brush or shake through again. As it lies in a heap, level or smooth it with the blade of a table knife, or any straight-edge; if thoroughly mixed it will present a uniform colour; if it appears darker in one part than in another, it must be sifted again. A sieve with a top and receiver is very desirable, as nearly all mixtures are either black or poisonous; the dust from star mixtures is very injurious to the lungs. If a top and receiver cannot be readily purchased, both may easily be constructed out of a sheet of millboard, fastened with a bradawl and waxed yellow flax, and neatly covered with paper. Small quantities of composition may be mixed up so thoroughly, in the mortar with the sash brush, as scarcely to require any sifting at all. The numerous bristles, as they are swept round, divide the composition so effectually and mingle it again so intimately, that the desired end is attained with the greatest rapidity. As much of the dust becomes entangled in the brush, hold it upright, with the bristles downwards, and give it a few raps at the top with a mallet.

Mixtures may be damped on a Dutch-tile, a marble slab, or a slate without a frame. They may be stirred about with a dessert knife, pressed flat and chopped, or minced as it were, and again pressed flat. Or they may be carefully rubbed up in a mortar with a pestle; which method, with the least solution, damps them most effectually.

To Make Lac Solution (shellac).

Put 1/2 ounce of flake shellac into a tin pot and pour upon it a quarter of a pint, or 5 ounces, of denatured or wood alcohol. Let it stand for about a day, stirring it occasionally till dissolved. Then half fill a basin with boiling water; set the tin, containing the lac, in it, and leave it till it boils and curdles. If the water does not remain hot long enough to make it boil, set it in a second basin of boiling water. Or it may be set in the saucepan, after being removed from the fire. As soon as it has curdled, remove it; and, when cold, pour it into a vial and cork it. Alcohol must never be boiled over a fire, nor near one, as the vapour will enflame. Keep the pot, therefore, while in the water, at a distance from a fire, or flame of a lamp or candle. Hot alcohol dissolves gum-resins more completely than cold; and renders the solution more adhesive. If the solution should, at any time, be thinned by adding more alcohol, it must be boiled again, or it will not mix properly.

To Make Wax Solution.

Put into a vial 1/2 an ounce of white wax (bleached beeswax), pour upon it 5 ounces of mineral naphtha (coal or gas tar naphtha), keep it tightly corked.

To Make Stearine Solution.

Dissolve a piece of a stearine candle in mineral naphtha, in the same way: suet will answer. Mineral naphtha must not be used near a candle or fire, as it gives off an inflammable vapour, at less than 100 degrees F.

To Make Gum Solution.

There is no better way of preparing this than simply to put cold water upon gum arabic and let it stand till dissolved. If for sticking purposes, as much water as will just cover the gum will be sufficient; but, for damping stars, or for making quickmatch, dissolve 1¼ ounce of gum in a pint of water. If required in a hurry, put the gum into *cold* water in a tin saucepan, set it on the fire, boil and keep stirring till dissolved. The vessel must be of sufficient capacity, as it is apt to boil over. When cold, bottle and cork it. Before mixing it with stars, care should be taken that it has not become sour; if it has, make a fresh solution. Acidity, above all things, must be avoided with mixtures containing potassium chlorate and sulfur. Test the solution with litmus paper. Gum solution is best made with cold water and not boiled.

To Make Dextrine Solution.

Take half an ounce of dextrine, and 5 ounces, or a quarter of a pint of cold water; put the dextrine into a cup or basin, add a little of the water, and mix it well with a teaspoon, rubbing it till all is dissolved; then add the remainder of the water, stir well together a second time, pour it into a vial, and cork for use. Dextrine, wetted to the consistency of honey, may be used instead of thick gum-arabic water for pasting. For this purpose it is advisable to keep either the gum or the dextrine in a wide-mouthed bottle, and to set the bottle in a gallipot containing a little water; the brush, a camel-hair pencil, or very small sash brush, with 1/3 of the bristles cut away on each side to render it flat, can then be kept in the water when not in use; this will prevent it, on the one hand, from becoming dry and hard; and, on the other, from getting clogged and swollen. It can be squeezed between the thumb and fingers, when wanted for use. The flat gum brushes now sold, bound with tin, are not pleasant to use, as the tin oxidizes and turns of a disagreeable brown colour. If there is a difficulty in obtaining a graduated water measure, one sufficiently correct for pyrotechnic purposes may be made with a vial. Paste a narrow strip of paper up the outside of the vial, weigh 4 ounces of water in a cup in the scales: pour it into the vial, mark the height, and divide it into 4 equal parts, for ounces; it can then be graduated into half and quarter ounces, and increased, if large enough, to 5 or more ounces.

To Make Paste.

Paste is most economically made in a zinc pot, which may be 4" deep and 3½" diameter. Put into it 2 oz. of wheat flour, add a little cold water, rub the two together with a sash brush till smooth and free from lumps, pour in more water till the pot is full within about 1", set the pot in half a saucepanful of water, put it on the fire, make the water boil, and keep it and the paste boiling for 2 or 3 minutes, stirring the paste the while with the sash brush. Remove it from the fire and allow it to cool. The paste is to remain in the zinc pot, in which it will keep good for a length of time, and beautifully white. Should it become watery in damp weather, stir in a little fresh flour and reboil it. Bend up a triangular piece of perforated zinc into a trivet for the pot to stand upon, to allow the water to flow between it and the bottom of the saucepan or a brass curtain-ring will answer.

Some recommend alum in paste; I think it best avoided, especially in cases intended to receive coloured fires. Alum is a double salt, a sulfate of aluminum and potassium; it has an acid reaction; and, coming in contact with potassium chlorate and sulfur or sugar, may cause spontaneous combustion. A drop of sulfuric acid instantly ignites stars containing them. At theatres, the clown sometimes fires a cannon with what appears to be a red hot poker, but which, in reality, is only a piece of wood painted red. A mixture is made of potassium chlorate and sulfur, or sugar, a glass bead is filled with sulfuric acid and the hole stopped up with shellac. This is laid in the

202

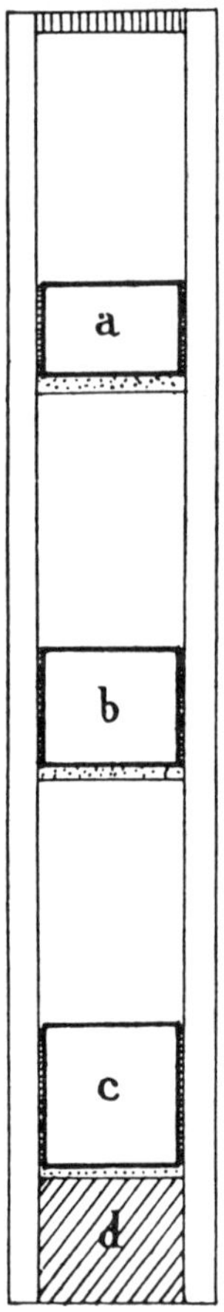

mixture; and when it is struck with the poker, the liquid escapes and enflames the composition. Copper sulfate is a particularly dangerous salt and must never be used, as it is almost certain to cause spontaneous combustion. Chertier, to whom pyrotechny otherwise owes so much, introduced an empirical preparation, by dissolving copper sulfate in water, together with potassium chlorate, drying it, and wetting it with ammonia: but this, however dried, when again wetted, turns litmus paper red. It has been named Chertier's copper. As safe and perfect blues can be obtained without it, it is best not to use it.

Two paste brushes will be sufficient for an amateur; sash brushs, one about 1" diameter, the other smaller, for light purposes. Let them stand in the paste. If they get dry, the bristles fall out. For convenience, one may be kept in the paste and one in water. Paste ought to be kept stirred occasionally.

Dry clay, powdered and sifted as fine as possible, is used for plugging, or stopping up the bottoms of cases. I have, for some time, discontinued its use, and employ plaster of Paris in preference. Directions will be given for each, so that the learner can adopt which he pleases; but plaster is infinitely preferable. Its use is an American improvement. Plaster of Paris (gypsum, or calcium sulfate), increases its volume when combined with water, that is, expands on crystallizing, so that it fills accurately every part of the case and leave no fissures; moreover, it gives no fire on being bored with steel, an advantage not always possessed by clay, or aluminum silicate.

Roman-Candle Scoops.

256

No species of fireworks require greater care in their construction than Roman candles. In the first place the stars must be fierce, that they may light thoroughly: next, they must not be driven out with too great velocity. For this purpose the blowing-powder must be carefully adjusted. The stars, also, must be of so easy a fit that when put into the case they may fall to the proper depth of their own accord. If they require pushing, they are too tight and will probably be blown out blind. When made as directed, they will, necessarily, be of an easy fit, as they will be of the inner diameter of the brass tube, while the bore of the case is equal to its external diameter; see figs. 202 and 256.

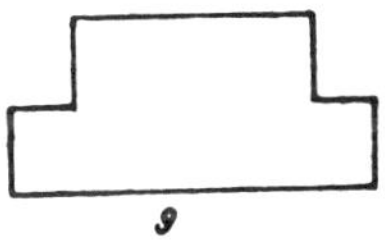

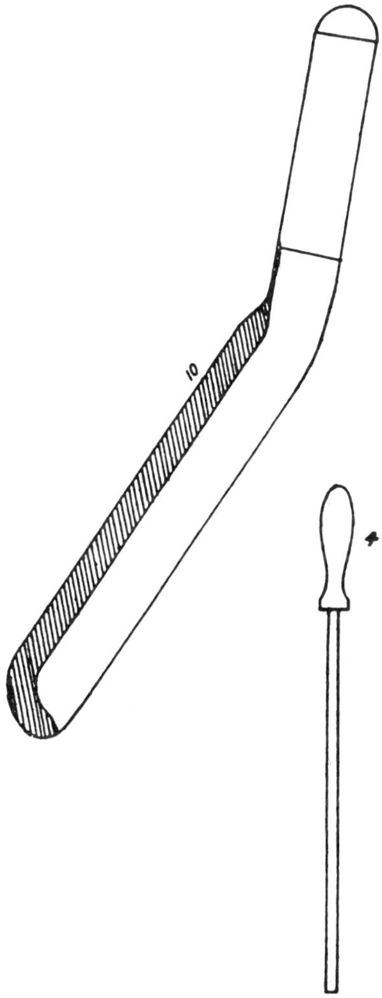

To regulate the blowing-powder, prepare a number of little scoops, like fig. 5, which is about the right size for the bottom star. They are formed of pieces of tin, zinc, or copper. Cut a long strip of tin 5/16" wide; cut this across into 7 pieces of the following lengths, 1-5/8, 1-6/8, 1-7/8, 2, 2-1/8, and 3". Round off the corners. Take a piece of brass wire, or stair-rod, about ¼" diameter, and with the wooden mallet, before mentioned, fig. 26, or with a hammer, bend each of the pieces round the rod into a half cylinder, or gutter. The best way to do this is to open the jaws of a vise about ¼" apart; lay the tin lengthwise flat over the aperture; hold the wire along the tin and hammer it down: this will drive the middle of the tin down between the jaws, and curl the edges up, giving it the semicylindrical shape at once. Having prepared the 7 strips in this way, take up the smallest, and hold 5/8" of the end of the stair-rod in the end of the semicylinder, to keep it open; put the other part from a to b, fig. 5, in a vise and pinch it up; it will assume the form represented; the bowl part will be 5/8" long, and the handle 1" long. Make the bowl of the next scoop 6/8" long, the next 7/8", and so on; the handle will always be 1" long. The last, for the top star, will have a bowl of 2". The smallest scoop ought to hold as much grain-powder as will weigh about 1/12 of the star; but to have the scoops accurate, it will be necessary to charge a Roman candle, fire it, and observe whether the stars go a uniform height; if they are blown out too violently and blind, two or three smaller scoops must be made, to adapt them to the strength of the powder and the fit of the star.

For measuring what is called the dark fire, or interval fuse, viz., the fuse between the top of one star and the bottom of the next, a large scoop, fig. 10, will be required. The tin may be 7/8" wide, and the bowl part 2½" long, bent round a ½" brass tube. To adjust it, take a Roman candle case, fit it on the foot, fig. 9, which is a cylindrical piece of wood, or brass, turned with a tenon to fit tight in the bottom of the case. fill the scoop and strike it level with a straight-cdgc, or with the rammer, fig. 4; empty it into the case, rest the foot on a flat surface, insert the rammer and jolt it gently up and down half a dozen times or more, lifting it about ½" at a time: put in another scoopful and jolt it in like manner. If the 2 scoopfuls, thus slightly compressed, fill 1" of the case, the scoop will be correct. If more or less the scoop must be shortened or lengthened accordingly.

A piece of writing-paper may be pasted and wound twice round the handle of each scoop, as from a to b, fig. 5. One dot, or stripe, or the fig. 1, can be put upon the scoop for the first or bottom star; two dots, or stripes, or the fig 2. for the second scoop, etc.; or any memorandum can be written upon them for future guidance. Should they get soiled, they may be cleaned with a soaped damp piece of sponge. A soiled leaf of a book may be cleaned in the same way.

For getting articles out of bottles, a teaspoon makes a convenient scoop; or the bowl part of fig. 10 may be fastened to a long wooden handle with a couple of tintacks. For small experiments, toy teaspoons are handy.

Gunpowder for fireworks is used in two forms; mealpowder and grainpowder. Mealpowder is a fine black dust, and is employed in all cases of mixing. Grainpowder is of three kinds, F, ff, and ffF; F, double F, and treble F; fine, double fine and treble fine. ffF is best for crackers, simply because it runs rapidly down the pipes; for driving stars, shells, etc., F will be sufficient; but ffF may be employed; ff need not be purchased. Very coarse powder is used in the trade for maroons, shells, etc., but is not necessary for amateurs.

If, in any place, there should be a difficulty in obtaining mealpowder, F grainpowder may be crushed in a leather bag, by laying the bag on a hard surface and beating it with a hammer, it must then be sifted through a very fine sieve; the leather should be of the same kind as shoes are made of.

To Charge Roman-Candle Cases.

Pour some F or ffF grain powder into a wooden bowl, or platter, fig. 11. Round the edge lay the little blowing-powder scoops, side by side, beginning with the smallest at a, the next at b, and so on to g.

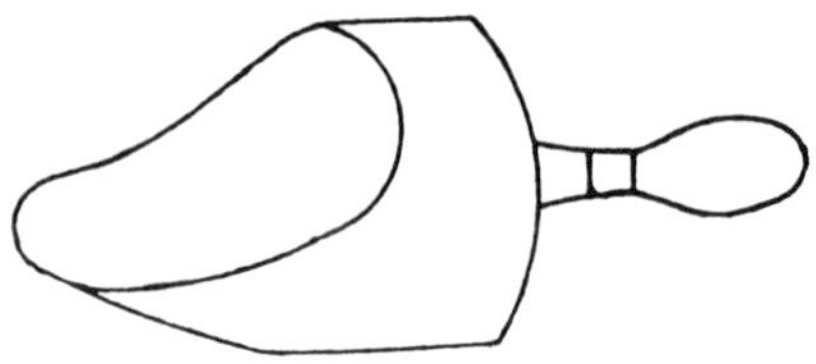

12

Put some Roman-candle fuse into a large tin scoop, made to stand on a flat bottom, like fig. 12, the same in shape as used by tea dealers; and, on the *right-hand* of it, lay the charging, or interval-fuse scoop, fig. 10. If the Roman candle is to contain different-coloured stars, set 7 in a row, in the order desired. When the cases are intended to be fired in threes or fours, the stars in one may be all blue, in another yellow, in another red, in another green.

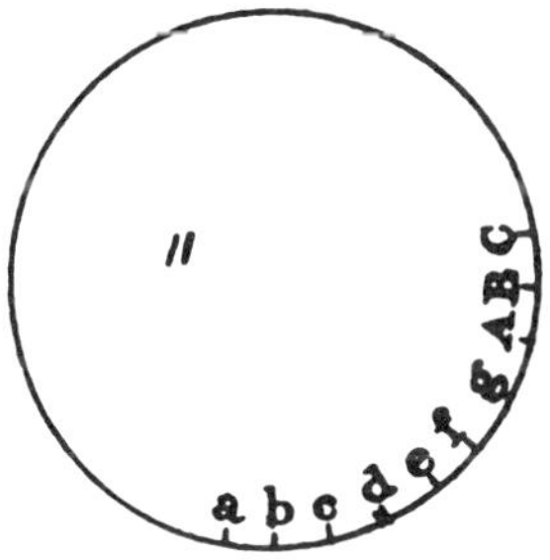

By the side of the stars have a cup containing some inflamant, and with it a scoop about the size of fig. 5. For distinction's sake, this may be formed of a quill; cut it to the shape of the bowl part of fig. 5, and plug the barrel part with plaster of Paris, pushed in with a wire. Now fit the cylindrical foot, fig. 9, in the bottom of the Roman-candle case; put in a scoopful, fig. 10, of powdered clay, insert the rammer, fig. 4, and jolt it firmly, till the clay is well compressed; let it fill up about ½", as shown at d, fig. 202. This being done, invert it and shake out any little dust that may remain. Put in the little scoopful, a, of the F or ffF grainpowder, of fig. 11, then lay the scoop at A. Now put in a star. As previously stated, it ought to drop of its own accord; but make sure that it has reached the blowing-powder, by putting in the rammer. Having ascertained this, put in the quill scoopful of the inflamant, and turn the case round, sloping, and tap it gently, to make the powder fall round the star. Now put in a level, or stricken, scoopful of Roman-candle fuse, with fig. 10, turning the scoop round, bowl downwards, when in, to insure the composition falling; withdraw the scoop and lay it on the *left* of fig. 12; insert the rammer, fig. 4, and jolt it gently up and down a few times, lifting it about ½" at a time; put in another scoopful of fuse, with fig. 10; lay the scoop on the *right* of fig. 12; insert the rammer, and jolt as before. Then proceed with the scoop b of grainpowder, and lay it at B; and so on, till the case is filled:

blowing-powder, star, inflamant, fuse, fuse. Repeat.

The fuse on the top star is best simply pressed in, a little at a time, as it is difficult to jolt the rammer in so small a space, besides that it puffs the dust up. The last 1/8", e, fig. 202, should be dry meal powder, forced in, by inverting the case and pressing it into a small heap, laid upon a sheet of paper; it binds better than Roman-candle fuse, and also blows off the leader pipe.

The blowing-powder scoops having been laid at A, B, etc., all that is required is to turn the bowl, or platter, a little round to the left, and they will come in rotation, ready for the next case; also, by putting the scoop, fig. 10, alternately to the left and right of the scoop, fig. 12, it will always be known whether the proper quantity of fuse has been put in.

Coloured stars, from their fierceness, have a tendency to burn in the cases; this defect is remedied by putting upon each star the

202

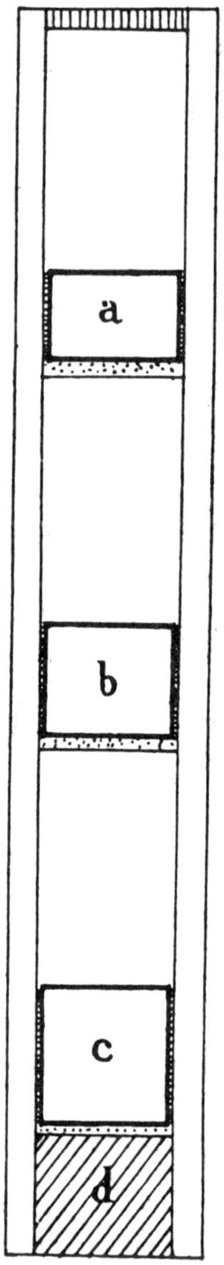

small quill scoopful of inflamant, which is sufficient to fall round the sides, and to fill up about the height of the star; at the same time the blowing-powder must be carefully adjusted, or the star, after it has caught, will be blown out blind from the too great rapidity with which it is driven. As stated, 1/12 the weight of the star will be required for the blowing-powder at bottom, if the star fits loosely; if it fits tight, 1/30 of its weight of grain powder will often be sufficient.

A Roman candle is well charged when the stars isochronise, or come out at equal intervals of time; they should also, theoretically, ascend to equal heights; but, with coloured stars, this cannot be perfectly insured, as some shrink more than others in drying, and, of course, fit more loosely; some are heavier, some fiercer, than others.

The interval fuse is best put in at twice, not at once, to insure more uniform compression; and especial care must be taken that it be not driven in too hard, otherwise the inflamant gets jammed so tight round the star that it burns *with* the star instead of blowing through, and the latter is not ejected till it is half burnt out. The quantity of composition that, driven firmly in a wheel-case, would rise only 3/4", ought to rise, in a Roman-candle case, 1" at least. Nothing spoils a Roman candle so much as hard ramming.

If a star should still burn a little in the case, the defect may be remedied by making it more squat, as b, fig. 202, or still more, as a. Otherwise the deeper the stars are, the longer they burn and the higher they ascend. Stars to be fierce enough for Roman-candles, must contain sulfur, or a sulfuret; without either, they are liable to miss: shellac, in powder, is excellent for matched rocket and shell stars, and renders them very durable; but it is not a perfect substitute for sulfur. The following will suit for Roman-candles; sugar blue, 1, 4, 5, 6, 7; blue, 1, 2, 5, 6, 13, 14, 17; yellow, 4, 8, 12; red, 1, 6, 7, 15, 16, 24; green, 1, 2, 7, 8, 9, 10, 17, 20; purple, 6, 14; mauve, 3, 8.

9

The interval fuse for all sizes of Roman-candles above 4/8 may be 1", as shown at fig. 202; for smaller sizes, 3/4" will be sufficient. The charcoal in Roman-candle fuse should not be too fine.

Instead of driving in clay at the bottom, plaster of Paris may be used, and then the foot, fig. 9, will not be required. Have some plaster of Paris in a wide-mouthed bottle; a glass of cold water with a salt-spoon in it; and a number of pieces of paper, about 4" square. Put a small quantity of the plaster on one of the pieces of paper; indent the middle with the finger; put to it a little water and work it up with a dessert-knife. Just as it gets to the consistency of mortar and is about to set, mould it with the fingers to the shape of a cork; push it into the end of the case; rest the case on a flat surface; insert the rammer and give it two or three slight jolts; turn it round a few times and withdraw it. Or press the plaster into the bottom of the case with the knife, set the case upright, empty a small scoopful, fig. 5, of Roman-candle fuse into the mouth; insert the rammer and jolt as before; the dry will prevent the other sticking to the rammer. Invert the case and shake out any dust.

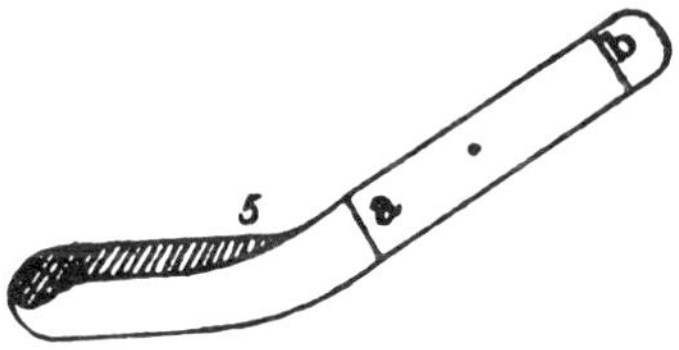

No more plaster must be mixed at a time than will suffice for one case. When plaster has once set, it cannot be mixed up a second time; therefore take a fresh piece of paper, and let the knife be cleaned every time. It is advisable to have two dessert-knives, then one can be used with which to scrape the other. As much plaster as will fill the case up about ¼" will be sufficient. They must be set by to dry; their not requiring the use of the foot will be found a great convenience.

Roman-candles are usually made from 3/8 to 6/8; but 5/8 is a very satisfactory size. If a Roman-candle is intended to be fired singly, twist a piece of touchpaper round the mouth. If the cases are intended to be fired in threes, fours, etc., to form a bouquet, or to be placed round a mine, jack-in-the-box, or devil-among-the-tailors, omit the touchpaper, and envelop the case in kraft paper, made to project an inch beyond the mouth, to receive the leader, or quickmatch. Roman-candle cases charged with tailed, streamer, steel, or comet stars, are called Italian Streamers.

A steel-pen inserted, nib backwards, in a penholder, makes a neat little scoop.

Cracks in wooden bowls may be stopped with plaster of Paris; and, if painted over with linseed oil, after getting dry, will remain waterproof for a long time. A screw may be made to hold in brickwork, by drilling a hole in the brick and pushing in the screw covered with plaster. Minute fissures, or defects in various articles, as of furniture, may be stopped with the same material, and blacked over to escape detection, or the plaster may be wetted with a mixture of black and red ink to match mahogany.

To Make Touchpaper.

Dissolve ½ an ounce of potassium nitrate in ½ a pint of hot water and let it stand till cold. Procure some 12 lb. colored tissue paper; cut each sheet into four equal parts, 15 by 10. Lay them smooth upon each other, and, with a sash brush dipped into the potassium nitrate solution, wash them over on one side and suspend them on a line to dry; afterwards string them together and hang them up for use.

To Make Slowmatch.

Dissolve 1 dram of lead nitrate in ½ an ounce of boiling water. Cut a sheet of blotting paper into six equal parts, and wet them on both sides with a sash brush, with the solution. When dry, paste a piece all over, and upon it smoothly press another piece; upon this, pasted, put a third piece; and so on, till all the six form a stiff board. Lay them under a heavy weight; and, when dry, with a sharp knife and straight-edge, cut the whole into strips ¼" broad. Four inches will burn about a quarter of an hour. Narrow tape, boiled in the solution, makes excellent slowmatch. Hold a piece, while burning, over a sheet of paper; the lead will melt and fall upon it in minute sphericles.

To Make Quickmatch.

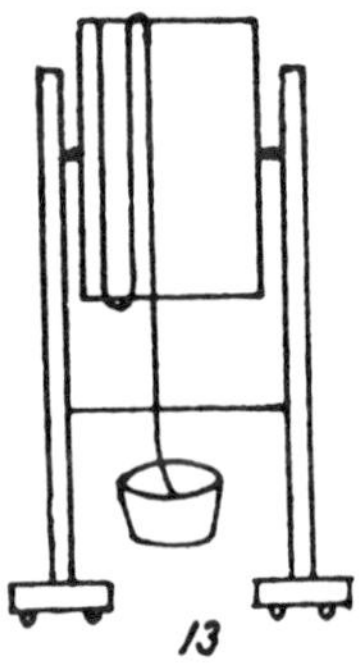

13

Put into a pan 2 lbs. of grainpowder, or mealpowder; pour upon it some thin hot starch, and stir it well about, breaking all lumps, till the mixture is of the consistency of paint. Procure some lamp-cotton, such as forms the wicks of candles. It will probably consist of sixteen or twenty-four strands. Divide it carefully into lengths of eight strands. This is not so easy a task as might appear. The best way is to act the reverse of a man spinning string. Divide the end of the cotton, say of sixteen strands, into two of eight each; fasten them to two screw-hooks a few inches apart. Take one in each hand and walk backwards, gently pulling them apart, and when they catch, untwist them: with care they will separate without breaking or entangling. Drop the end of one of the pieces into the pan, and, as it keeps falling, coil it round and round in the mixture, and press it down with an iron spoon, until as much is pressed in as the quantity can saturate. Be very particular that it is thoroughly soaked. Have ready a wooden frame, fig. 13, of pine, resembling a swing looking-glass, with the glass taken out. It may be 5 feet by 4. The frame is to be supported on pivots between two uprights. Fix a nail or hook at the left-hand corner of the frame and tie the end of the cotton to it which has been hanging outside the pan. Get a person to slowly turn the frame and hold it steady. Take hold of the cotton in the right hand, shut the hand, and allow the cotton to slip gradually and slowly through it, as the frame is turned; squeeze it very gently, so as to allow it to come out well coated, and contrive to make it as round as possible. When all the cotton is wound upon the frame, spread some sheets of paper--old newspapers--on the floor; at each corner place a brick; lay the four corners of the frame upon the four bricks: sift dry meal-powder all over the match, turn it over and do the other side. Prop the frame against the wall and let it dry. One ounce of white starch will be sufficient for a pint and a half of water. Rub the starch up smooth with a little of the water, then add the rest and boil it. Use with the 2 lbs. of grainpowder.

A New Method.

232

Put into some hot starch, made as before directed, or some cold gum-water, or cold dextrine solution; and with a small stencil-brush, or a 3/8" sash brush, cut across the middle to make the bristles short and stubby, stir in some mealpowder, till it is well mixed and of the consistency of pudding. To avoid repetition, it will be convenient to refer to this under the name of mealpaste.

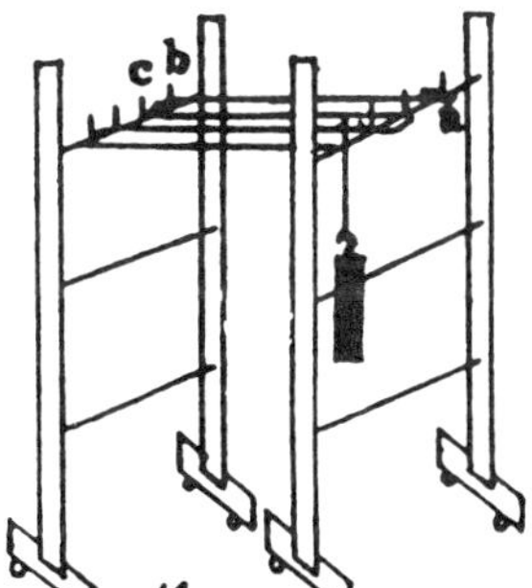

14

15

Take 2 towel-horses, fig. 14, and set them parallel, 5, 10, or 20 feet apart. In the top rail of each, drive four nails ½" asunder. fix the dry cotton to the nail a, carry it across to the opposite nail b, pass it round the second nail c, bring it to the opposite second nail, and so on, till the cotton lies in four parallel lines, like the strings of a harp placed horizontally. Hitch the cotton, without cutting it, to the hook of the weight, fig. 15. This weight is made by taking a piece of brass tube, 1" diameter and 4" long; stop one end with a bung; fill the other part with melted lead; when cold, pick out the bung, and set it with the vacant space upwards. Take a ceiling hook, fig. 232, hold it with a pair of pliers upright in the middle of the vacant space, and pour in more melted lead till full. Now procure two pieces of planed pine board; one 4" square, the other 6" square. With the left hand hold the smaller piece close underneath the cotton; and with the sash brush or stencil-brush, work the meal-paste *well into the fibres of the threads,* pressing the cotton on the board till it is thoroughly soaked, and rolling it over, laterally, to make it as round and smooth as possible. It is best to begin on the left and work towards the right. If the wetting slackens the threads, pull them tight. Now brush some mealpowder through a fine sieve, to free it from lumps; put a tablespoonful or two upon the larger board; hold it close under the four threads, as the other, with the left hand, and move it laterally forwards and backwards and down the whole length, at the same time brushing the meal over the threads, with a soft, dry sash brush, till they are smothered, and giving them an occasional jar to shake off the superfluous meal. By a little practice they may be made as smooth and as round as a piece of wire. Leave the weight hanging to them till they are dry. Instead of four nails on each rail, a dozen may be put; and if the towel-horses are set 12 yards apart, 144 yards may be soon made. In this case, three or four heavier weights would be required. They are unnecessary for an amateur; four hooks or nails opposite four others will be sufficient; and 6' or 8' apart is a good distance. A small quantity may thus be made one day and a small quantity another; and for this purpose it is best to use gum-water, as it is always ready, and a little can be added to the dry left the day before and a little fresh meal stirred in. Three or four threads of white darning-cotton, which is of two twists, make very good match; knitting or crochet cotton, which is of three twists, produces, alone, excellent match; two or three pieces of the first, or two of the latter, put into a leader pipe, side by side, blow through with a violent report. Match may also be made of the very narrowest white tape, 3/16" wide; this, from its flatness, is peculiarly suited for enveloped stars and lance-work. For general purposes, however, lamp cotton is decidedly the best, as it is most loosely twisted; and, therefore, the most absorbent. If match is liable to be exposed to damp, it should be cased in thick leaders and be prepared with starch; in other respects, nothing can surpass gumwater; and, to a beginner, the easiest way of all is to make only one piece of match at a time. Cut the cotton of such length as may be convenient, tie a knot at each end to prevent the separation or entanglement of the threads; fasten one end to a nail, and let the other end hang over a rail or chairback; and, instead of putting the dry meal on a piece of board, have a piece of tin, 8" by 4", and bend it into a gutter 8" long, and lay the powder in it: or half a sheet of note paper can be held in the hand, and bent into a semicylinder; the match can then be easily smothered. Tape, from its numerous interstices, is so absorbent that it does not require to have the meal-paste rubbed into it with the stencil brush. Having stretched a piece tight with the weight, hold it steady with the left hand, and paint 4" or 5", first on one side, then on the other, with meal-paste, with a camel-hair pencil cut across; then 4" or 5" more, and so on. It may then be smothered with dry meal, or instead, it may be left to dry, and then receive a second coat of meal-paste. It is best to begin on the right and work backwards, as the weight pulls against the hand and keeps the tape steady.

Match ought when cut across, to look black throughout; it should be stiff, straight and round--or flat, if tape--but, to test it, cut off about 10", put one end of this into a leader pipe, so that 5" will be in the pipe and 5" out; hold the end of the leader with a pair of tongs, or lay it on the ground; light the naked end. If the match is good it will burn gradually, though swiftly, till it gets to the pipe; it will then blow through with a bang. The nearer the leader pipe fits the match the better, only it ought to be large enough to allow it to go easily in without force: if laid in a Roman-candle case it would hardly puff: the smaller and smaller the tube becomes the louder and louder the report: the increase of power being in the inverse ratio of the diminution of space. A train of gunpowder laid in the open air, or confined in a tube, comports itself in the same way, as is well known. It explodes instantaneously, because the gas cannot escape as quickly as it is formed, but, passing through the tube, ignites the whole. Match, when dry, should be kept straight; for, if it gets broken, it acts like a cracker, snapping at every break.

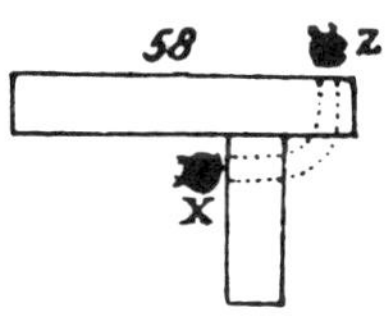

A piece of zinc rainwater pipe, of suitable length, furnished with a bottom and lid, or a couple of bungs, is convenient for keeping it in; but a far better contrivance is a pine box, 5 feet long, 3" deep and 3" wide, made of ½" pine. The lid of this box can be readily furnished with three or five hinges, like a piano, only made with string instead of brass. Take two pieces of string, fig. 57, and tie them in a knot near the end, as at k. At x, fig. 58, make a bradawl hole through the back of the box, near the top, and push the strings through it; make another hole through the lid, put the strings through it, and tie in a knot on the top, as at z: and so with the other four. This will render the explanation easier to understand; but, practically, a single piece of thicker string is best, if you can manage to tie the second knot in the exact spot you wish. The knots must not be too close to each other, as the lid requires a little play. For safety, the box ought to be furnished with a lock; and, in this case, 4 brass hinges; but I give the string method, as it is convenient to know it.

ROCKETS.

CHARGING HEAVY ROCKETS

Rockets are charged in choked cases on a spindle, to leave a hollow up the middle through which the fire may be communicated to nearly the whole of the composition at once: this causes the sudden generation of an enormous quantity of elastic vapour, which, being unable to escape instantly through the contracted aperture left for its exit, exerts its pressure in the contrary direction and hurries the rocket forward. A stick, attached to it, guides it like the rudder of a vessel, or tail of a bird, or fish; while its weight and leverage keep the centre of gravity a little below the case, and prevent the rocket from pitching over. Its manufacture, therefore, from the commencement of cutting the paper for the case, to its finish of fitting on the stick, requires an accurate adjustment of all its parts. Disregarding the trade names of pound, ounce, etc., which, now that moulds are dispensed with, are useful only for enabling the makers, from tradition, to understand each other; the internal diameter, instead of the external, as formerly, may be selected from which to compute the relative measures. Taking the bore of the case as unity, the proportions will be as follows:--

1	Internal diameter of case.
1½	External diameter.
8	Length of case.
6	Length of spindle.
2/5	Bottom diameter of spindle.
64	Length of stick.
½ by ½	Thickness and breadth of stick.

These proportions are most readily calculated by taking the diameter in eighths of an inch.

Selecting, for instance, a 5/8 rocket, we have

6/8, inner diameter;

half as much again,

9/8, outer diameter.

Taking the numerator as inches (which is multiplying by 8);

6", length of case;

3/4 of this, or, which is the same, ½ the outer diameter, as inches,

4½", length of spindle; putting 20 for the denominator, instead of 8 (8 being 2/5 of 20),

6/20", bottom diameter of spindle;

6X8, (numerator multiplied by denominator),

48", length of stick;
half by half,
3/8 by 3/8, size of stick.

The following table exhibits the usual sizes in inches.

Name of Rocket	Inner Diameter	Outer Diameter	Length of Case	Length of Spindle	Bottom dia. of Spindle	Length of Stick	Size of Stick
10/8	10/8	15/8	10	7½	10/20	10 times 8 = 80	5/8 by 5/8
9/8	9/8	13½/8		6¾	9/20	9 times 8 = 72	4½/8 by 4½/8
8/8	8/8	12/8		6	8/20	8 times 8 = 64	4/8 by 4/8
7/8	7/8	10½/8	7	5¼	7/20	7 times 8 = 56	3½/8 by 3½/8
6/8	6/8	9/8	6	4½	6/20	6 times 8 = 48	3/8 by 3/8
5/8	5/8	7½/8	5	3¾	5/20	5 times 8 = 40	2½/8 by 2½/8
4½/8	4½/8	6¾/8	4½	3-3/8	4½/20	4½ times 8 = 36	2¼\8 by 2¼/8
4/8	4/8	6/8	4	3	4/20	4 times 8 = 32	2/8 by 2/8
3/8	3/8	4¼/8		2¼	3/20	3 times 8 = 24	1½/8 by 1½/8

To Make a 6/8 Rocket.

6

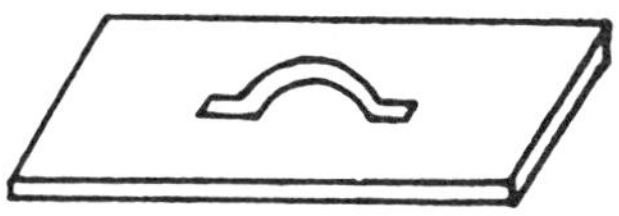

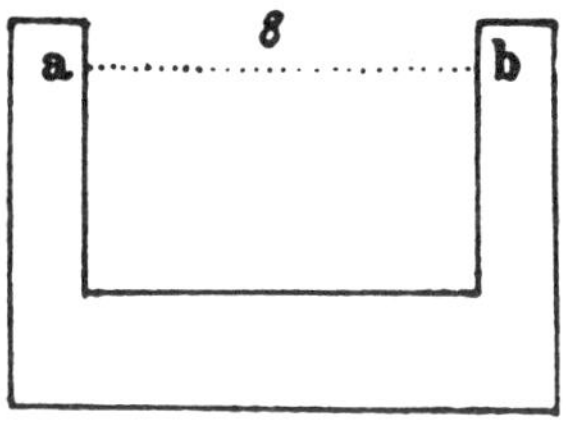

Have the former, fig. 7, a brass tube 6/8" external diameter; the gauge, fig. 8, with a 9/8 aperture. Procure some kraft paper, 70 or 80 lb., the thicker and heavier the better. The smooth machine-made cylinder-dried answers very well. The sort used for laying under carpets, wood-pulp paper, and which is 5 feet wide, is almost equal to the original rough kraft, and should be procured, if possible, in preference to the smooth. Cut it into strips 6" wide. Paste the pieces well, and roll the cases as hard as possible with the rolling board, fig. 6, till they fit the gauge. Lay them by for a few hours to get partially dry. They must then be choked, or strangled, about half a diameter from one end of the case, that end that lay nearest to the left hand in rolling, till they assume the shape of the neck of a vial. For this purpose have a cylindrical piece of pine, alder, or any kind of wood, fig. 24, about 9" or 10" long, made to fit easily into the case; cut it into 2 pieces, a and b; b may be 1½" long; round off the ends just cut. In a, fasten a screw, the solid wire, or uncut part of which, is ¼" thick; then saw or file off the head, and slightly taper the part projecting; in the piece b bore a hole to just fit the wire of a. Slip b upon a, and push them into the case, so that the interval between the two reaches within about ½" of one end of the case; slightly pull out b, as drawn, to leave a neck or hollow, round which to form the choke; fix a staple or screw-eye in a post; tie to it one end of a piece of cord about a yard long, and 1/10" thick; fasten the other end round the middle of the stick for a handle; take hold of the stick with the right hand; hold the case in the left; pass the cord round the space left vacant, and gradually tighten it, turning the case round and round, so as to pinch it in equally on all

22

23

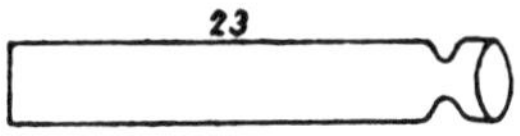

16

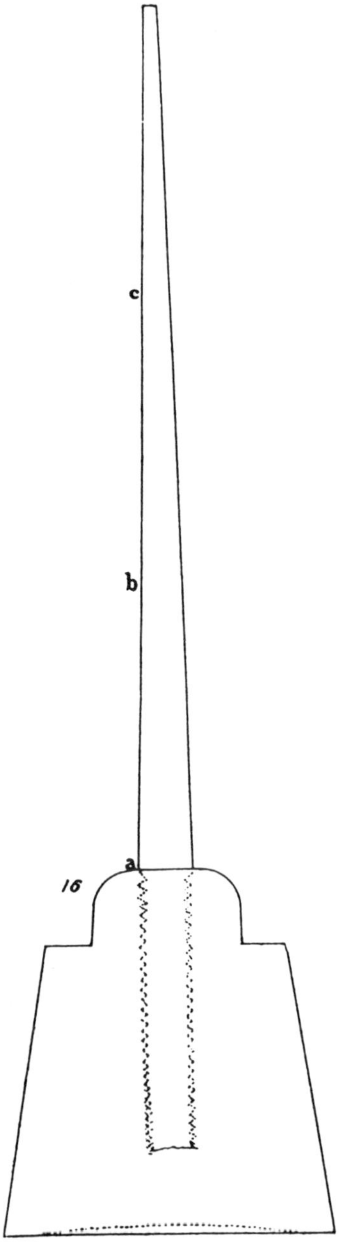

sides, till it assumes the shape of fig. 23. Hitch a piece of string, about 1/16" thick, a few times round the neck, until it is filled up flush with the other part of the case, remove the choker, and hang the case up to dry. A hitch is made by simply bending the string backwards into a loop, like fig. 22; passing it round the choke; pulling it tight; looping it again, again passing it round the choke, and so on. The first loop is called a half hitch, and will not hold of itself; the second loop completes the hitch, and makes it hold; the remaining hitches are for filling up the choke, and restoring the cylindrical shape. Practice, by hitching the string upon your left thumb, 4 or 5 times; it will hold after the second loop; but push it off at the end, and it will all fall loose again. The best kind of string for choking is laid cord; whipcord is too tightly twisted, and snaps; as also does cable-laid cord.

A case must not be choked when wet, or it will tear; nor must it be too dry, or it will be difficult to choke it at all. The drier, however, it can be choked the better. Experience is the only guide. If properly dry, the wrinkles of the choke will be small and perfectly regular. Should the string stick to the case, chalk it; but this is not likely to occur unless the paper is too wet. Have a foot, fig. 16, turned of hardwood, with a hemispherical nipple 6/8" diameter, as drawn. Bore it with a twist drill to the depth of $\frac{1}{2}$". Procure a brass, iron, or steel wire, 6" long and 3/10" diameter, and perfectly straight. Cut on one end of it a screw, 1$\frac{1}{2}$" long, fix it in a vise, smear the screw with glue or with thin plaster of Paris, which will rapidly rust it and cause it to bite; then screw the foot on. File the 4$\frac{1}{2}$" gradually tapering. The object of tapering it is simply to make it deliver. A cylindrical hollow up the rocket would answer as well, but the spindle could not be got out. The more conically true it is tapered, of course, the better. Finish it off with a very fine file, and smooth it with glass paper or emery cloth. The block into which it is screwed may be larger than drawn; the bottom should be turned slightly concave, to make it stand firm, on the same principle as the bottoms of plates, cups, etc., are made with a rim. The block and spindle are better if case in brass, or gun-metal, in one solid piece; the pattern to give to the founder should be in one solid piece of wood. After casting, the rough parts must be filed smooth; the spindle is better if turned in the lathe.

The next articles required are a setting-down piece, fig. 17; three hollow rammers, or drifts, fig. 18, 19 and 20; and one solid rammer,

17 18 19 20 21

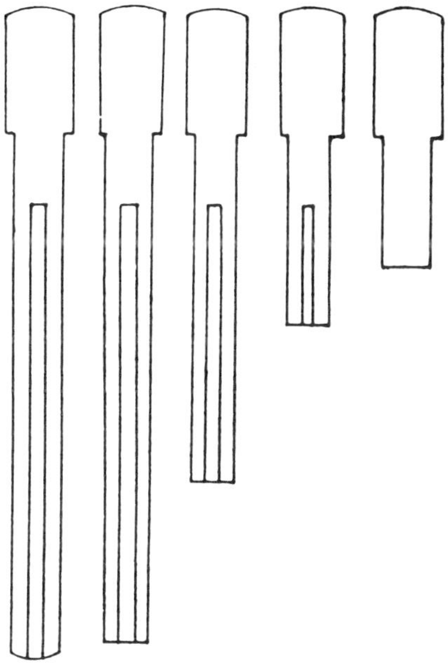

fig. 21. They are simply cylindrical pieces of wood, turned with a head to bear the blows of the mallet. Hardwood, free from knots, will answer. The lengths of figs. 18, 19, 20 and 21, are to be 8, 6, 4 and 2 diameters respectively, exclusive of the head; that is, 6, $4\frac{1}{2}$, 3 and $1\frac{1}{2}$". The hollow drifts are to have a cylindrical hole bored up them to within one inch of the handle, so as to clear the spindle by $\frac{1}{2}$", to allow of any dust being driven up it. The hollow in figs. 17 and 18 must be large enough to fit the spindle loosely at a; in fig. 19, a trifle smaller, to fit loosely at b; and in fig. 20, a trifle smaller still, to fit loosely at c.

The next requisite is a hardwood mallet, fig. 25. It may be a cylinder, 5" long and 3" diameter, with a handle about 5" long and 1" diameter, turned out of the solid. Let it be turned at the end slightly concave, like the bottom of fig. 16, that it may be set to stand upright like a wine bottle. Or the head may be made 3" square, like fig. 26; or a small carpenter's-mallet, about a pound and a quarter weight, will answer. A pair of cloth gloves will prevent the hands from getting chafed, while charging cases.

To Charge Rocket Cases.

The first requisite is a solid block of wood, 6" or 8" square, and 18" or 20" high. This is indispensable. A piece of old oak gate post answers well. It must be set upon the ground, or on a flagstone, not on a floor. It is impossible to ram a rocket properly on a floor because of the vibration. It is, also, necessary to sit, not stand, at the work.

25

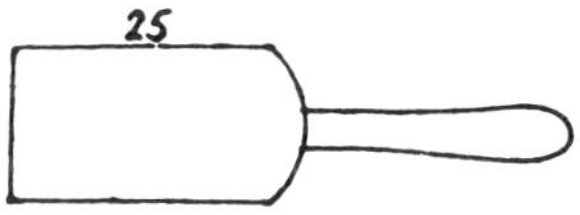

26

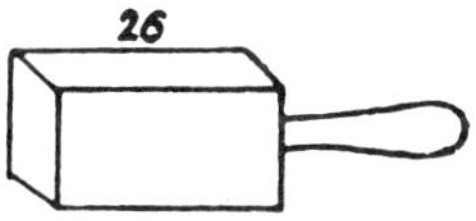

Put the setting-down piece, fig. 17, into the case; press the case over the spindle, and give the head a few blows with the mallet; this will smooth out the wrinkles of the choke, which is all that fig. 17 is used for. Now put in a very little powdered clay, and mallet it with fig. 18; as much clay as will reach up 1/12" will be sufficient; its object is to preserve the choke from burning and getting enlarged. One cannot be too precautious with rockets. Now put in a scoop of rocket fuse, with fig. 10, insert drift, fig. 18, and mallet the fuse in firm with about a dozen and a half blows, or till it offers a resistance to the hand. The blows must be light and numerous, not slow and heavy, like driving a post into the ground. Eighteen blows, with a momentum of 3, will consolidate the fuse: 3 violent blows with a momentum of 18, would perhaps bend the case, or drive the dust up into your eyes. Shift the drift round occasionally. The mallet need not be lifted about 4" or 5" at a time. If the rocket is not rammed firm throughout, it will, upon lighting, explode.

As soon as the case is charged about $1\frac{1}{2}$", make a pencil or ink mark, round the drift, where it stands level with the top of the case, for future guidance; then charge another $1\frac{1}{2}$" with the second rammer, fig. 19, and mark it in like manner; proceed in the same way with fig. 20. It is obvious that if fig. 19 were used too soon it would get split by the spindle being driven up it, and the spindle would be bent or broken, hence the advisability of marking the drifts, to know when to lay aside one and take the other. Just before you get to the top of the spindle, put in the solid rammer to feel how high the spindle reaches near the top of the case; hold it by the thumb and finger to keep the distance, and mark it down the outside of the case by indenting the case with the edge of the drift. Exactly 1-1/3 diameter, that is, in this case, exactly 1", above this indentation make another mark; then, as soon as you have covered the spindle till you can no longer see it, with the use of fig. 20, begin charging with the solid drift, fig. 21, till the composition inside is level with the top mark outside. This being done, drive in a little dry clay till the case is full. Remove the rocket from the spindle by giving it a turn or two round to the right, not backwards; then bore a hole through the clay till you can see the composition, with a 3/16" bit. The bit should be fixed in a handle, and kept for the purpose. It is not advisable to use a brace and bit unless the bit is shielded, as it is apt to bore too deeply. The bit may be fixed in a handle by boring a large hole in the handle and pouring in melted lead, or pressing in plaster of Paris.

215

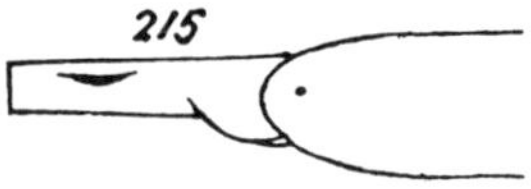

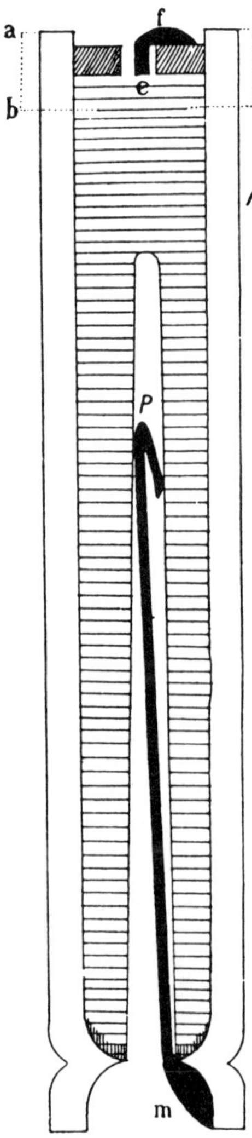

194

Instead of driving in dry clay on the top of the composition, for a plug, a little plaster of Paris may be pressed in, and a hole pushed through the middle with an oiled bradawl. As soon as the plaster is dry, the hole can be enlarged and bored perfectly clean; whereas the clay is apt to crumble, and chip out. A good instrument to clear out the hole with is a penknife, broken across, and ground to a chisel shape, fig. 215. The object of the clay, or plaster, is to prevent the composition, which, containing much charcoal, does not bind well, from getting disturbed, and the solid part diminished, which would cause the stars to be ignited while the rocket was ascending, or the fuse, perhaps, to blow through at the beginning. For asteroid, or parachute rockets, a bent piece of naked quickmatch, ef, fig. 194, is to be inserted in the hole, through the clay or plaster, and retained in position by a dab of wetted meal pressed on the end f. For ordinary rockets, into the heads of which the bursting powder is put loose, this precaution is not necessary, and the hole may be left open. In all rockets a long piece of match, p m, is to be bent to a hook to make it catch, and pushed up the core and fastened to the side of the mouth with a little dab of wetted meal-powder, m, pressed on it with the blade of a knife.

36

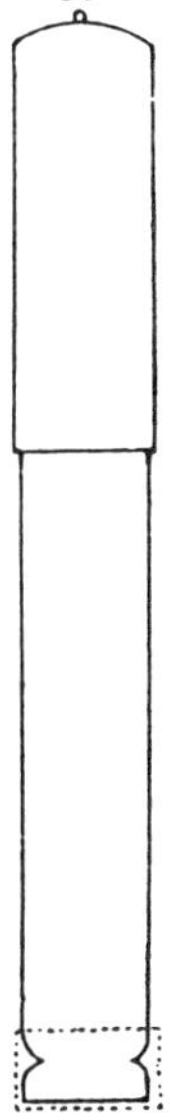

If the rocket is intended to be lit with a port-fire, take a circular piece of touch-paper, about 2" diameter, slightly paste it all over, lay it in the left hand, press the mouth of the rocket down upon it and smooth the edges of the touch-paper up round the case. Otherwise, smear the end of the case with the sash brush dipped into mealpaste, and when dry, paste a bit of touch-paper round it and twist to a point, like a squib. The appearance of the rocket complete is shown at fig. 36; the dotted line shows the touch-paper.

In driving with the hollow rammers, it generally happens that a little of the powder gets driven up the hole; this, if allowed to accumulate, is very troublesome to remove; it should, therefore, be knocked out every time, by holding the drift in the left hand, and giving the head a rap or two with the mallet.

The whole of the composition ought to be put in, in about as many scoops as the length of the case contains half-inches; thus 6 scoopfuls ought to fill a 3/8 rocket; 8 scoopfuls a 4/8 rocket, etc. Make a scoop for each-sized rocket, and write upon the handle the name of the rocket it belongs to. These directions may appear minute, but they will save much trouble if attended to.

As it is convenient to know about what quantity of composition will be required for any rocket, the following formulas will be useful:--

$$E/9 = \text{Drams avoirdupois; } 3E = \text{grains troy.}$$

Where E denotes the size of the rocket, in eights of an inch:
Required the quantity of powder for a 6/8 rocket.

6x6x6/9 = 4 x 6 - 24 drams = $1\frac{1}{2}$ ounce;

3 x 6 x 6 x 6 = 648 grains.

For a 3/8 rocket?

3 x 3 x 3/9 = 3 drams.

3 x 3 x 3 x 3 = 81 grains.

The same weights denote the quantity of stars which the rockets will safely carry: thus an ounce and a half of stars may be put into the head of a 6/8 rocket, and 3 drams into a 3/8, avoirdupois; or about 648 grains troy into a 6/8, and 81 into a 3/8. Along with the stars is to be put in from 1/10 to 1/12 their weight of bursting powder. The following table will serve as a guide:

Size of Rocket	Weight of Comp. or Stars	
3/8"	3 drams	or 81 grains troy
4/8"	7 drams	or 192 grains
4½/8"	10 drams	or 273 grains
5/8"	14 drams	or 375 grains
6/8"	1½ oz.	or 648 grains
7/8"	2-3/8 oz.	or 1030 grains
8/8"	3½ oz.	or 1536 grains
9/8"	5 oz.	or 2187 grains
10/8"	7 oz.	or 3000 grains

$3^3 : 3 :: 4^3 : 7 :: 5^3 : 14 : 6^3 : 24.$

The rockets will actually carry more stars then indicated, but it is best not to overweight them.

With respect to the quantity of powder required, the table is intended as a guide for only about its weight, to ascertain the quantity for filling a certain number of rockets. Suppose it were desired to charge a dozen 4/8 rockets, with powder No. 1, what quantity would it be advisable to mix?

7 x 12 = 84 drams = 5¼ ounces required.

The proportions are, 1 + 4 + 8 = 13; therefore, halve the numbers, and take

½ oz. sulfur + 2 oz. charcoal + 4 oz. nitre = 6½ oz. powder.

Suppose it required to try a 3/8" rocket with powder No. 14. The numbers are 1 + 2 + 4 + 2 + 1 = 10; and 81 grains troy are required. As steel is heavy, rather more must be mixed, so multiply each by 10, and we have, 10 grains sulfur + 20 charcoal + 40 nitre + 20 meal + 10 steel = 100 grains. The bursting powder may be measured with the scoop used in charging the rocket; for a 5/8" and 6/8", use a stricken, or level, scoopful; for smaller rockets, rather less than a scoopful; for larger rockets, a heaped scoopful; or, 1/10 may be taken, and so the last, or unit, figure omitted from the weight given in grains; thus for a 3/8" take 8 grains; for a 4/8", 19 grains; for a 6/8", 64 grains; but it is not necessary to be very particular. Should any difference be made, take rather less rather than rather more.

70

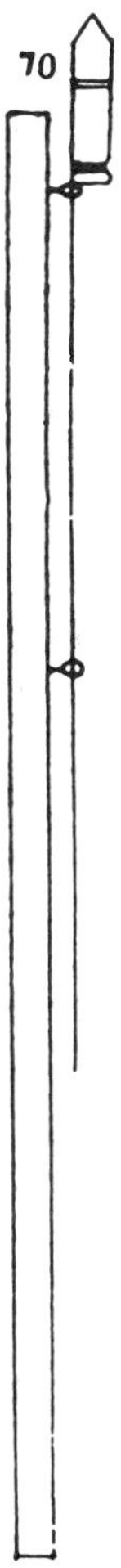

Rockets should be fired from 2 staples, or 2 screw-eyes, fixed in a post, one near the top, the other half a yard below, as in fig. 70; or they may be suspended on a nail near the top and passed through a staple, or screw-eye, below, which mode prevents the stick from dragging. They should never be propped against a wall, a chair-back, a gate, or railing, as they might fall, especially on a windy night, after the touch-paper was lit and before the fuse had caught. Every possible care should be taken in guiding them, as it is too late to think about any mischief they may cause after they have once started.

In making rockets, it is essential, above all things, to have good potassium nitrate and charcoal. The best way with fresh materials is to weigh out as much nitatre, charcoal and sulfur as will make one small rocket. Have the nitrate as fine as possible, and dry it over the fire in the 6-inch frying-pan. If the rocket ascends well, you will know that the articles are pure, and you can proceed to use them; but if the rocket does not rise, you may conclude that the articles are adulterated, the nitrate with salt, or that the charcoal is, perhaps, merely pine sawdust burnt in a retort. In this case you must buy the nitrate in crystals, and the charcoal in sticks. To powder the nitrate, put it in a pipkin, pour on it a little water, set it on the fire, make the water boil, and keep stirring with a piece of wood until it is dry and of a fine powder. Charcoal is best ground up in a coffee-mill. It must first be broken into small pieces about the size of coffee-beans. After being ground it should be sifted through a sieve, the meshes of which are about 1/40" square. To get a correct idea of this size, lay down a line on paper 1" long; mark it off into eight equal parts by taking the divisions from a foot rule; then carefully divide one of the eights into five equal parts. But the best way is to make a square pine sieve, about 8" square and 3" deep then nail on the bottom a piece of perforated zinc, with ¼" Flemish tacks. The perforations are circular, and should be 20 to the inch, measured diagonally; that is, a diamond shaped inch, or inch rhombus, consisting of 20 oblique rows, each row containing 20 holes; 20 x 20 = 400. These holes will be about the same size as the ones mentioned in the sieve, because the solid parts between the perforations are also about 1/40". A strip of pine, ¼" thick, should be nailed round the bottom to keep the zinc tight. A square box for a receiver, and another for a lid, should also be constructed, otherwise you will be smothered in sifting charcoal. Making use of the perforated zinc sieve just described, all the charcoal that goes through, fine and coarse, should be used for rockets. It need not be shaken much, but brushed through with a sash brush. If the coffee-mill be properly set, it will grind it of the requisite fineness, at once. Construct a second sieve with 15 perforations to the lineal inch. Iron borings for gerbes should be sifted through this; use all that goes through. Construct, also, a third sieve with 9 perforations to the inch. Nothing can equal iron and steel for making sparks; but neither will keep long from rusting after coming in contact with nitrates. Coke grains, about the size of pinheads, are a fair substitute. Beat coke into lumps about as large as peas, then grind it in a coffee-mill and brush it through the 9-perforation sieve; sift the fine dust from what passes through; throw the dust away and keep the grains. Porcelain may be powdered in an iron mortar, for gerbes; brush through the 9-perforation sieve; sift out the fine dust with a fine sieve; throw it away and keep the grains. The intense heat of the focus of the choke renders them incandescent; and, from their weight, they are projected to a considerable distance. They are inferior to coke grains, as the latter are to steel filings.

Before grinding a fresh substance in a coffee-mill, it must be taken to pieces, brushed clean, and screwed up again. Fine lawn or hair sieves should be used for sifting chemicals; excellent sieves may be made with book-muslin; the cylinders may be 4" diameter, 3" deep; the muslin should be cut into a circular form, and hemmed round a piece of string; it may then be slipped over the drum, or cylinder, and secured with an elastic band; or it may be pasted up and round the sides, and if about 4" diameter, two pieces of string, or tape, may be crossed over the middle of the bottom to strengthen it.

Charcoal may be made by putting some dry pieces of willow, alder, poplar, sycamore, maple, or almost any kind of wood, except the pine or turpentine tribe, into an old iron saucepan, covering them with perfectly dry sand, and setting the saucepan in the middle of a fire, to remain red hot till the wood is completely burnt through and smoke ceases to escape. Remove when judged sufficiently charred; and when cold, not before, pour away the sand. A new kind of charcoal is now made of straw.

Sulfur is used in the state of sublimed sulfur, sulfur sublimatum, or flowers of sulfur, and when mixed with nitrate, requires no preparation; but as it is always more or less impregnated with sulfuric acid, as is readily shown by testing it with litmus paper, it might, on coming into contact with potassium chlorate, cause spontaneous combustion. To prevent this, it is necessary to wash the sulfur. For this purpose put it into a pan, and pour upon it boiling water in which some salts of tartar (potassium carbonate) have been dissolved; stir it well and break down all lumps. Let it stand to subside; pour off the supernatant liquor; fill up with cold water and let it stand, to again subside. Make a conical bag, fig. 33, with a piece of linen or calico, sewed at the top round a ring or hoop of wire or cane, or whalebone; fasten a string to it, by which to hang it up. Put the washed sulfur into it, and hang it under a water tap; turn the water gently on, and let it drip all night; this will wash away every trace both of acid and alkali. Afterwards hang the bag up three or four days till the sulfur is dry; it may then be bottled, and kept exclusively for colours. Roll brimstone, crushed and sifted fine, is suitable for stars at once, as it is free from the sulfurous acid generated by burning the sulfur in oxygen during the process of sublimation; but powdering and sifting it is a very troublesome operation.

33

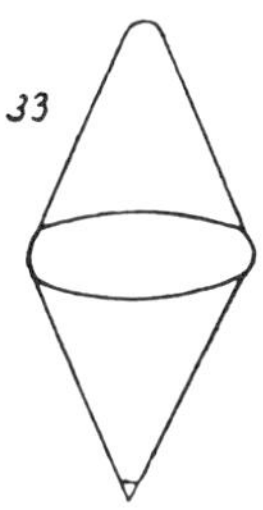

Sodium oxalate may be made thus--procure 3 lbs. of sodium carbonate, the common washing soda used by the laundress, not bicarbonate of soda; boil it up in a saucepan with just as little water as will suffice to dissolve it, which will be about half a pint. Dissolve, in another vessel, 1 lb. of oxalic acid in boiling water, and pour it into a large bowl, capable of holding two or three quarts. Now put to this the dissolved sodium carbonate with a tablespoon, a spoonful at a time. A violent effervescence takes place. The soda is to be slowly added till effervescence ceases. It should be tested with a strip of litmus paper to see if the acid is perfectly neutralized.

Copper oxychloride may be made by laying thin pieces of copper in a dish, and pouring upon them a mixture of half water and half hydrochloric or muriatic acid. The next day remove them and lay them on a board in the shade to dry. When dry brush off the green powder which will be found on the outside, with a tooth-brush, into a basin of water. After a quantity is obtained, wash it as directed for sulfur, and dry it in the bag, fig. 33. Test it with litmus paper to ascertain if free from acid.

To prepare litmus paper, dissolve ¼ of an ounce of litmus in an ounce and a half of hot water; when thoroughly dissolved, and the water is of a dark blue colour, take some writing paper, and with a sash brush or camel-hair pencil, go over it on both sides with the litmus solution. When dry, wet some of these prepared pieces, with the brush dipped into vinegar; this will turn them red. Dry and preserve both. They may be cut into strips, ½" wide; the blue strips will be tests for acids; the red for alkalies. Wet a strip of the blue and touch it with oxalic acid, it will turn red; wet a strip of the red, or the piece just reddened, with sodium carbonate, it will turn blue. Litmus books can be bought for 2*d.* each.

To make copper disulfide, procure some thin sheet copper about as thick as a card; cut it into pieces, and put it into a crucible with sulfur, a layer of sulfur and a layer of copper alternately, till full. Set the crucible in a clear fire, and keep it red hot for an hour. Remove it; when cold, break it up in an iron mortar, or grind it in a coffee-mill. Sift in a lawn or book-muslin sieve as fine as possible. Half a pound of copper and a quarter of a pound of sulfur may be employed.

There is a black sulfide or sulfuret of copper produced by passing sulfuretted hydrogen through a solution of protoxide of copper: this is useless, and also dangerous.

For want of a coffee-mill, charcoal may be beaten in a leather bag, with a hammer.

A variety of rocket powders will be found in the Tables; the first is as good as any, and will answer for all sizes from 3/8 to 12/8. As a rule, the fiercer powders, containing mealpowder, may be used for small rockets, but are by no means necessary. Try Nos. 11, 14 and 18.

A rocket, when starting, makes a roar; but this is not on account of the fierceness of the powder, but of the extent of the surface ignited. Rocket composition, laid in a train, burns very slowly: the finer the charcoal, however, the fiercer it will be: so to get a rocket to rise to the greatest elevation, have the charcoal impalpable: this will elongate the tail; and of course, at the same time attenuate it; as 4 times 8 are the same as 8 times 4.

ROCKET STARS

Rocket stars are made in three or four ways. First cut, or chopped or naked stars. This mode is used for nitrate stars only; potassium chlorate stars require different methods. It is a singular circumstance, that though potassium chlorate stars are much fiercer than nitrate stars, yet the latter light without any trouble, while the former, if made in the same way, would be almost sure to miss.

To Make Cut Stars.

Wet the composition with thin starch, or dextrine solution, or gum-water, sufficiently to bind; press it into a flat mass, on a slate, or Dutch tile, with a knife, or small trowel, till about 3/8" thick. Indent the surface with the edge of the knife, in parallel lines about 3/8" apart, and cross these with equidistant indentations, at right angles. Set the mass by to dry gradually. When nearly dry, break it up into little 3/8" cubes, and lay them out to dry thoroughly. The broken edges will be rough and will catch easily.

Dry Pill-box stars.

Take a sheet of note paper and cut it into four equal parts; each part will be about 4½" by 3". Paste and roll them on a 7/16" brass tube, so as to have the cases 4½" long. To make these into pill-boxes, perfectly true, like those used by the druggist, they must be cut in the lathe. For this purpose, turn a cylindrical piece of wood, fig. 27, which is to fit easily into the case, except at **a**, where it is to be turned sloping a little larger, so that when the case is slipped over it, it will bite at the part **a**; otherwise, on putting the chisel to it, it would slip round without getting cut. Mark the case, with a black-lead pencil, at every half inch; suspend it in the lathe, and cut it at the marks. Next procure a 7/16" punch, with which to cut out the bottoms. These are to be made of card, or Bristol board. Lay the card on a piece of sheet lead, or the grain end of a piece of beech; hold the punch upright on the card, and give it a smart blow or two with a hammer; keep on punching, till the punch contains a dozen or more discs; then push them up with a piece of wood, or cane, and they will all fall out from the middle, which is enlarged for the purpose. When a number are ready, press them into the pill-boxes with a rammer that fits loosely. The composition is to be put into these dry, and driven in with a solid rammer and the little mallet before described.

The above is needlessly exact for fireworks. The following method will be found preferable. Cut writing paper, or 30 lb. cartridge paper, or 50 lb. kraft paper, 4" by $3\frac{1}{2}$"; paste it, and roll it on a 7/16" brass tube, so as to have the case 4" long; when dry, with a strong pair of scissors, clip it sharp across the middle; it can readily be guessed at. Clip each of the two pieces across the middle; clip each of the four pieces across the middle; there will then be 8 boxes. The clipping slightly flattens the cases, but they can readily be restored to the cylindrical form by slipping half a dozen at a time on a piece of cane that nearly fits, and passing them under the rolling board, fig. 6. It is best to paste writing paper on both sides, before rolling, as it is apt, otherwise, to blister in drying. Should it be wished to indicate the colour of the star, a strip of thin double-crown, 4" long and 2" wide, may be pasted round the dry case, before it is cut into pill-boxes.

To Fill the Boxes with Dry Colour.

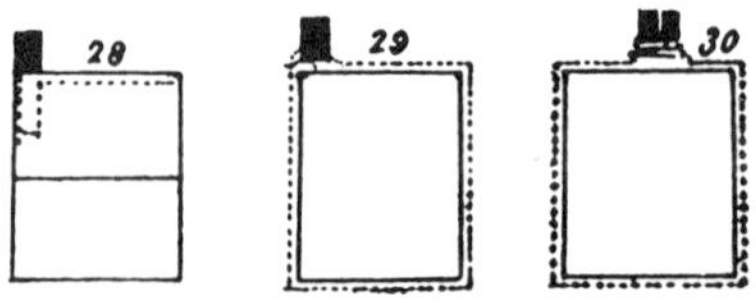

Rest the pill-box on a flat surface, put in some composition, and drive it in with a brass drift, till half full, as in fig. 28; then fill up the box with more colour, set a little bit of match upright in the side, and mallet it in, till the box is full. The drift should be filed flat up the side, about 1", so that a cross section of it will be of the shape of the capital letter D. Cut a piece of tissue paper, about 2" long and $1\frac{1}{4}$" wide; paste it; lay the pill-box on it, and roll it up; tuck in one end to make a bottom, and press the other end round the match, till it assumes the form of fig. 29.

Another way is to set the pill-box on a flat surface, put in a very little dry mealpowder, then some composition, and mallet it in till full. Roll this up in a piece of tissue paper as before; tuck in the bottom, and set by to dry; when dry, put into the other end--the end containing the film of meal--a piece or two of thin match and tie it in, as in fig. 30. This way, however, is almost too tedious for stars, and is better suited for case colours for wheels.

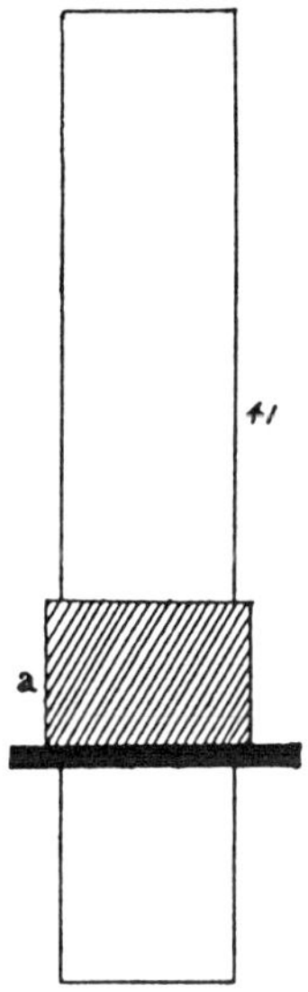

A quicker way than either of the preceding is the following. Cut a piece of thin tissue paper, $1\frac{1}{2}$" long, and 1" wide; paste it, and roll the pill-box in it, so that the paper shall be flush at one end and overhang the other 1/2"; tuck in the overhanging part to form a bottom, and set by to dry. When dry, mallet in composition till full within 1/16th", and so prepare a number. Rub up, in a mortar, a little chlorate mealpowder, with lac-solution; spread it out on the Dutch tile, or slate, and level it to about 1/8" thick; invert the pill-box, and press it into the damped priming, to fill up the 1/16th vacancy. Dry pill-boxes are best for making chameleon stars, or turncoats, as they are sometimes called; these are half one colour, and half another; make a dozen half-yellow and half-green; and a dozen more half-blue and half-red; put them into a rocket head; they will burst green and red and change to blue and yellow.

Bottomless Pill Boxes.

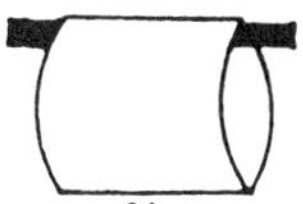
31

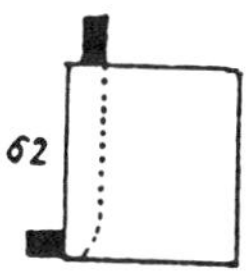
62

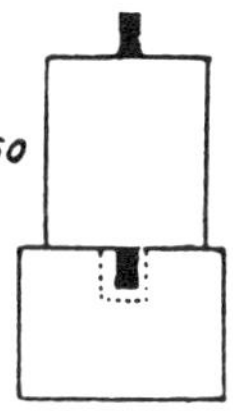
60

265

266

The cases are the same as before. To fill them, *damp* the composition, as for Roman candle stars; put a bit of quickmatch into the case, as at fig. 31, and press in the composition. This is usually done with the fingers, but it is not very pleasant work, especially with lac solution. A cleaner way though more tedious, is to fit the case on a little foot, with a side notch in it, fig. 60, then press in the composition with a drift the thickness of a blacklead pencil, or with a blacklead pencil. Or a notch may be cut in the side of the box, fig. 61; and the match put in, as in fig. 62, and the composition pressed in. An easy way is shown at figs. 265 and 266; the former with tape, the latter with 4-strand cotton match, cut into $1\frac{1}{2}$" lengths. Put a piece into the box and bend it, as drawn; it can then be held by the match, while the compositon is pressed in; the match being of such length is sure to catch.

Enveloped Stars.

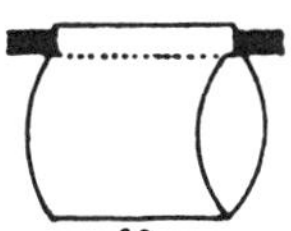
32

Pump and drive the stars exactly as for Roman candles: fig. 32 is a good size. Set them by to dry. Cut a strip of tissue paper, of the proper colour to indicate the star, $3\frac{1}{2}$" long and $\frac{1}{2}$ wide; paste it slightly all over, and lay the star upon it as at **a**, fig. 41, with a bit of match behind it, then roll it up, and lay it in a warm place to dry quickly, to prevent the absorption of the paste into the star; it will have the appearance of fig. 32. In fig. 31, the match is embedded, and dries in the damped composition; in the enveloped stars, the stars are dried first, and the match lies outside and blows through; they show well in the air. For getting the greatest number of stars

253

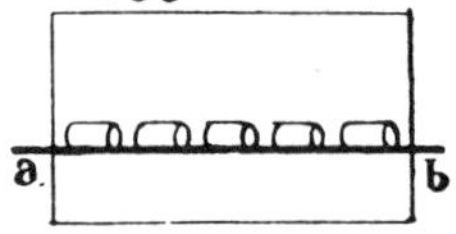

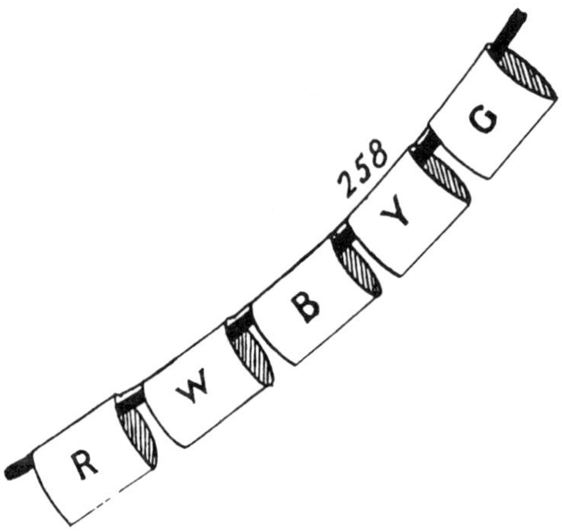

into the smallest space, the following plan will be found convenient. Cut a piece of tissue paper, 3" long, $2\frac{1}{2}$" wide, fig. 253; paste it slightly all over; lay a piece of match, 4" long, upon it, as **a b**; and, in front of it, 5 stars, previously thoroughly dried, about 1/8" apart; roll them up tight, and dry quickly. When dry, with a table knife cut the paper between the stars to detach or isolate them, and leave them hanging by the match only, as by a hinge, as **RWBYG** (red, white, etc.), fig. 258. Damped stars are preferable to dry pill-box stars. Too much care cannot be bestowed upon the preparation of coloured stars; it is more economical and satisfactory to put a dozen into a rocket head and have them all light, than to put in a score and have half of them miss. Compositions containing powdered shellac make durable rocket stars. Try also green 21 and 22; and red 23 and 24. A fierce star, matched and enveloped in a strip of pasted tinfoil, like fig. 32, with the match cut flush at each end, may be put into a Roman candle for the top star; it will need no special inflamant, and but little blowing-powder; it should be made a tight fit in the case, so as to require pushing down.

Naked Coloured Stars.

Select fierce, or sulfur compositions; as red, 1 or 6; green, 1 or 17; yellow, 8 or 18; blue, 10 or 17; damp the blue with stearine solution, or gum-water or a little of each; the red, green and yellow with wax solution. Pump, and drive them in a tube former 7/8" or 8/8" diameter; and make them $\frac{1}{4}$" thick, like large peppermint lozenges; when dry, break each into two; they will be found to be very vivid and durable, and, from their occupying little room, well suited for small rockets. Wax solution imparts durability, without impairing the inflammability, and gives variety of tint; if used with blues, it turns them into mauves; it modifies all colours, improving some and deteriorating others. Stars may be impregnated with stearine by damping them with stearine solution; the mixture may then be spread out to dry; afterwards damped with gum-water, pumped and driven. Benzine as well as mineral naphtha, is a solvent of stearine. Nitrate stars may be made in the peppermint lozenge shape, and broken into two, instead of being chopped.

Unmatched Stars.

Rub up the composition in a mortar, with lac-solution if containing strontium nitrate; with gum-water, if not; just sufficient to make it cohere; empty in on to a slab and slightly level it with the blade of a knife, to the depth of about half an inch. Take a pill-box case, open at both ends, set it upright in the composition; push in the drift, which may be a blacklead pencil, or wire of equal thickness; invert the box, set it down again; again push in the drift; and so keep on, inverting and pushing in the drift, till the box is full. If now you wish the star to light at one end only, cut a strip of thin tissue paper blue, and brush it over with a thick coat of gum-solution; set the pill-boxes upon it, press them down and leave them to dry. When dry, cut round them, and each will be furnished with a bottom. Now rub up some chlorate mealpowder with the least possible quantity of lac-solution, or gum-water, the same as used with the stars; empty it; level it to about 1/8" thick, and press the mouth of the pill-box upon it. If you desire it to light at both ends, you of course omit the bottom, and prime both ends. The entire surfaces then burn with immense rapidity, and show most vividly in the air.

Oiled Stars.

Oil will suit only a few mixtures; the following may be tried; red, No. 7; green, 17; yellow, 8; blue, 10 or 17; steel, 1. Use boiled linseed oil, which is linseed oil boiled with litharge, or oxide of lead, to free it from the mucilaginous and albuminous substances contained in the seeds of flax, and which retard its drying. Put into a mortar 1 oz. of composition; add to it 12 drops of oil, and rub it up with the pestle till thoroughly incorporated. Oil does not sink into a mixture like water or alcohol, but requires long rubbing like the preparation of putty. Press the mixture into pill-boxes, like the stars just described. The chlorate meal may be rubbed up with oil or lac-solution. Press the mouths into it, as before recommended.

Rocket Heads.

Heads for small rockets may be made of two or three rounds of paper rolled dry, and secured on the inner and outer edges with paste. After sticking it on the case, which it may be made to fit, as in fig. 36, pinch the top in like a choke, only tight, and tie it round with twine or flax. For coloured rocket heads, 6/8 and upwards, the head may be enlarged by fixing a collar round the top of the case. To make the collar for a 6/8, roll a case on a 9/8 former, and when dry, cut it into short lengths in the lathe as recommended for pill-boxes; if unprovided with a lathe, saw it with a fine-toothed saw; or a long strip of paper 1/2" wide, may be pasted and wound round the case. The advantage of the enlarged head is, that it brings the stars nearer to the rocket, and prevents it from being top-heavy. Another way is to make the heads tapering, and the tops conical, as

35

39

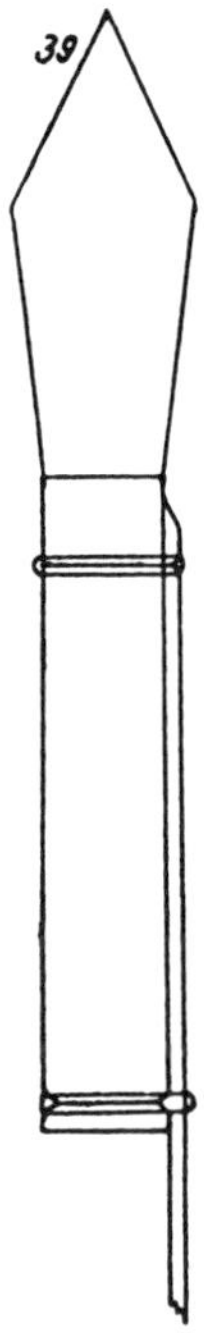

34

37

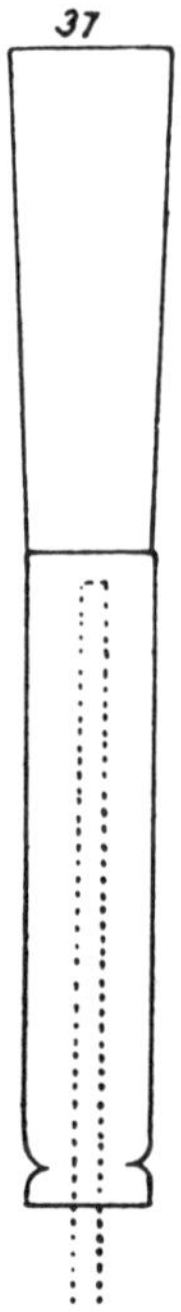

38

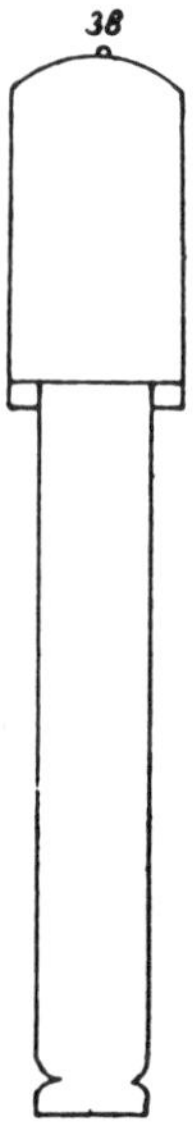

in fig. 39. One part is rolled on fig. 34, the other on fig. 35; or, a cone may be made of a circular piece of paper, without a former. Cut the circle, along the radius, to the centre; paste it all over, and bend it into a cone, pasted side outwards. When dry cut a larger circle of tissue paper; cut along the radius to the centre; paste it, and wind it round the cone, pasted side inwards; snip the edge with the scissors, then secure it to the rocket head.

But the quickest way of all is to make a long paper bag, which may be made to fit to the greatest nicety. For this purpose take a tape-measure; or lay down, on the edge of a strip of writing paper, 7 or 8" long, and 1" wide, 6", divided into eighths, transferred from a foot-rule. Suppose the 6/8 rocket has a collar, which, on being measured by the paper just alluded to, is found to be 4-2/8" round; add to this 3/8" for lapping over, making 4-5/8". Cut a piece of kraft paper 4-5/8" wide, and as long as the case, 6". Make this into a paper bag, 2-1/8" wide. Be careful that the corners are perfect; a strip of tissue paper may be pasted over them. When dry, pinch the mouth open till cylindrical, and merely allow the stars to drop in without forcing them; this will keep the top edge of a wedge shape, fig. 37, and answer the purpose of a cone. If the paper is thin, the bag must be made of two thicknesses; the paper will, then, require to be 9" by 6; 4-2/8 of this will have to be kept dry, and the other 4-6/8 pasted.

A head, made in this way, if required to hold rains, or serpents, can instantly be reduced to a cylindrical shape, by pushing the rocket-case right up it, to the top; this will cause the corners to stick out, like two horns; press them down, and secure them with sealing-wax. Attach a cone, if desired.

Heads made like figs. 38 and 39 may be of three thicknesses of paper, pasted all over.

Fig. 39 shows the manner of tying on the sticks. No variation must be made in their lengths, and it is not advisable to alter their size. To adapt it to the wood, however, a slight alteration might be permitted. For instance, instead of 3/8" square, it might be 7/16" by 5/16", a slight increase one way, compensated by a slight diminution the other. It must, however, on no account, be so increased and diminished, as to approach the shape of a lath, as such stick would vibrate, and cause the rocket to quiver. When the heads are a paper bag of the shape of fig. 37, the stick must be tied on, as indicated by the dotted lines.

40

If the stick is suited to the rocket, it will, when suspended on the finger almost against the mouth, as at f, fig. 40, lie, not quite horizontal, but slightly sloping downward. The wood should be dry pine, free from knots. The sticks are generally cut with a carpenter's cutting-gauge. If the learner has a lathe he will find a 6" circular saw convenient for cutting them.

221

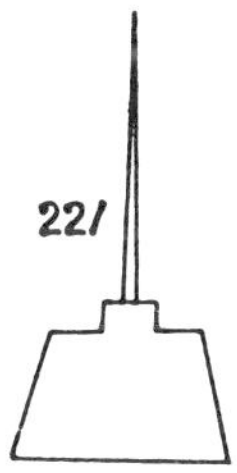

For amateurs, a 6/8 rocket is a good size; large enough, and small enough. If a 5/8 and 4/8 are made, two hollow drifts will be sufficient; for 3/8, one hollow drift. Very small rockets, 2/8, are made for children; they are rammed solid, and a hole is pushed up them with a bradawl: the powder may be No. 14 or 18. Rockets may be made with clay vents without any choke at all. Have the spindle screwed into a foot, flat at the top, like fig. 221, let the foot enter the case $\frac{1}{4}$", to keep it steady; drive in $\frac{1}{4}$" of dry clay, as hard as possible; then begin charging the rocket with fig. 18; fig. 222 shows the charged case. Plaster of Paris can be used instead of clay; let it be driven on a foot like fig. 42, and cleared out with fig. 215. The foot must be smeared with tallow or sweet oil to prevent the plaster from adhering; then use fig. 16.

222

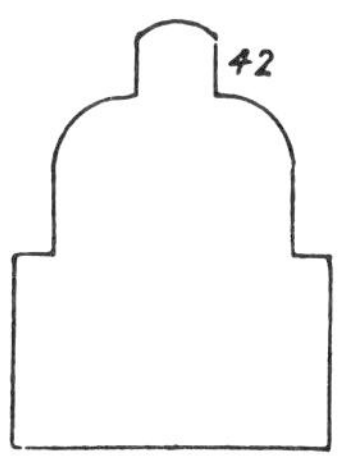

42

Rockets in former times, before the present days of competition, were charged in moulds. These were of gun-metal, bored truly cylindrical, furnished with hinges to open and admit the case; they were then screwed up, and might be charged as hard as possible. Names were given them according to the bore of the mould, that is, the external diameter of the cases; a 4/8" was termed an ounce rocket; a 5/8", a two ounce; a 6/8" a quarter pound; an 8/8" a half pound; a 10/8", a pound; a 12/8", a two pound.

These names were determined by the weight of a leaden ball of the same diameter as the bore. Now a sphere of lead, 7$\frac{1}{2}$" diameter = 90 lbs., or 1,440 ounces; consequently one of 15" = 720 lbs; similar solids being to each other as the cubes of their like dimensions: the latter sphere being twice the diameter from top to bottom; twice the diameter from left to right; and twice the diameter from front to back; 2 x 2 x 2 = 8.

The external diameter of the rocket being 7½", the internal would be 5", and this would be a 90-pounder: hence, for an inch rocket, we the the proportion,

$$\overset{\text{oz.}}{5} : \overset{\text{oz.}}{1{,}440} :: 1^3 : 11{\cdot}52$$

so that a 12-ounce, or 3-quarter-pound rocket, ought to be a trifle above an inch.

The names in use enable the makers to understand each other, but they are not mathematically correct, and are of no utility to an amateur.

The following table shows the true weight of leaden spheres, the dimensions being taken in inches.

Weight	Diameter	Weight	Diameter	Weight	Diameter
1 dram	.264	1 oz.	.664	1 lb.	1.672
2 dram	.332	2 oz.	.836	1¼ lb.	1.8
4 dram	.418	4 oz.	1.056	1½ lb.	1.91
8 dram	.528	8 oz.	1.328	1¾ lb.	2.02
12 dram	.604	12 oz.	1.52	2 lb.	2.112

And 2/3 of the above numbers multiplied by 8, give the correct names for rockets in eighths of an inch.

The bore of rockets, *as generally understood,* is exhibited in the following table.

½ oz.	3/8 inch	½ lb.	1 inch	7 pounder	2-1/7 inch
1 oz.	4/8 inch	1 lb.	1¼ inch	10 pounder	2-2/5 inch
2 oz.	5/8 inch	2 lb.	1½ inch	16 pounder	2-4/5 inch
4 oz.	6/8 inch	3 lb.	1-3/5 inch	20 pounder	3-1/8 inch
6 oz.	7/8 inch	4 lb.	1-4/5 inch	24 pounder	3¼ inch

WHEEL AND FIXED CASES.

Wheel cases may be 5-5/8" long. Cut the paper, without waste, into 4 strips, each 29" long. Fixed cases may be 7¼" long. Cut the paper the other way of the sheet, into 4 strips, each 22½" long. The extra length for fixed cases, is to allow of their being reported: that

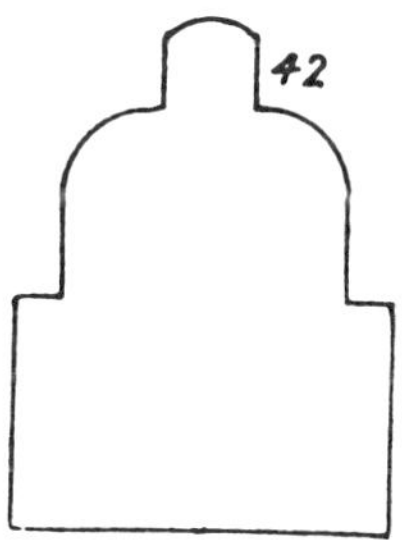

is, filled at the end, with an inch, or more, of grain powder, to make a bang like a squib. The cases are to be gauged to the thickness of Roman candle cases, but choked like rockets; 5/8" internal and 7/8" external are a good size. Two solid drifts will be required for wheel cases; one 5½" long, besides the head; the other 3"; for the fixed cases it will be necessary to have one of 7¼". They may be turned of box, beech or ash; but gun-metal drifts are best, though by no means indispensable. If they are cast in gun-metal, they will simply require filing in the rough spaces. The cases are to be rammed solid throughout, on a foot, fig. 42, turned in one piece, with a nipple, and a pin 1/3 of a diameter thick, and just high enough to keep the choke clear.

In all wheel and fixed cases, whatever the remainder of the powder may be, begin by putting in one scoop of starting-fire; this, when malleted firmly in, should fill about 3/8" in the case. The subsequent powder can be selected from the Tables. The choke may be protected with a little clay, before the starting fire, like rockets, if thought desirable; no clay is to be used anywhere else. Wheel cases containing steel filings are termed brilliant.

To prime the cases, very slightly damp some mealpowder, by sprinkling it with a few drops of water; mix and chop it up, or mince it, as it were, with a knife; put a little into the mouth of the case; press the nipple, fig. 42, into it, and work it round; this will prime the choke and the mouth at once, still leaving the choke clear; or paint the choke and mouth with the sash brush, with meal paste, just sufficient to wet them, then plunge them into dry meal, give them a rap to shake off the superfluous dust, and lay them by to dry. If the chokes have been protected with a film of clay, it will be necessary, of course, before priming them, to see that the throat is clear, so that the priming may reach the powder.

Wheel and fixed cases need not be choked at all, but plugged ¼" with plaster of Paris, as directed for Roman candles. When dry, charge them by setting them flat on the block, without using the foot. Mallet in a scoop of starting-fire, then the other. Afterwards, when a number are charged, bore through the centre of the plaster, with a shell-bit, 1/3 the diameter of the case; insert a piece of match in the hole; wash the face of the plaster over with meal paste, and plunge into dry meal. Or drive the plaster down on to the foot, fig. 42, and when dry, charge as if choked.

A sheet of tissue paper makes eight wheel-case envelopes (2.2.2) 10 by $7\frac{1}{2}$", so that the void at each end, to receive the match, is nearly 1".

It is not possible to devise a formula that will indicate the exact quantity required for wheel cases, saxons, tourbillions, etc., as the powders vary; but representing the length in inches by l; and the diameter in eights by **e**, the following will help to serve as a guide:

$$le^2/7 = \text{drams}$$

Suppose a wheel case 5-5/8" long and 4/8" diameter, what quantity of powder will be required?

$$5\text{-}5/8 \times 4 \times 4/7 = 13 \text{ drams}$$

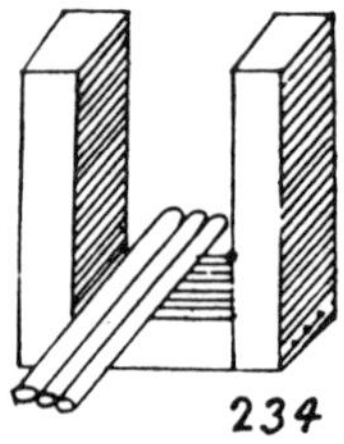

234

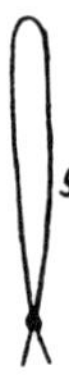

99

To construct a frame for holding the cases, take a piece of pine, 16" long, 4" wide and 1" thick; cut it into three pieces; two pieces, each 6" long, and the other piece 4" long. Nail or screw them to the trough shape of fig. 234. Cases may also be held together with a bit of elastic, tied in a knot, like fig. 99.

GERBES.

Gerbes, so called from the French word for wheat-sheaf, which they resemble, are fixed choked cases: they do not show well on wheels. As they contain grains of iron, they must not be under 6/8"; for private exhibitions a good size is 9/8", 9/8" internal diameter; 12/8" external; 3/8" the diameter of the choke; $11\frac{1}{4}$", length of case. Charge the case on a nipple, exactly like a wheel case. It is advisable to put in first a little clay to protect the choke, as the fire, being fierce, would, otherwise, enlarge it, and diminish the ascent of the sparks. Upon the clay drive in a scoop of starting-fire, and fill up with gerbe composition. This, when containing iron borings, is termed Chinese fire; the pieces, Chinese trees. The cascade at the Crystal Palace is made with gerbes of 3" bore, clay-vented.

The most magnificent of all, however, is the coloured gerbe, a golden sheaf with intermingled flowers. For this, some green, blue and red grains, or small stars, must be prepared. They may be $\frac{1}{4}$" cubes, cut as directed for the chopped nitrate stars. A far better way, however, of preparing them, is to remove the pin **a**, of fig. 2, up to **c**, so that the stars, driven in the tube, will be 5/8" diameter and 2/8" thick. When these are dry, chop them into four pieces, by holding a knife or chisel across them and giving it a smart blow with the mallet. They are harder made this way.

To charge the cases: having driven in the clay and the starting-fire, put in 7 or 8 stars, then a scoopful of powder No. 1 or 2 or 6, then 7 or 8 more stars, and another scoop of powder; mallet the whole 4 layers down firm, with blows not too heavy at a time, but many times repeated. Then put in 7 or 8 more stars, another scoopful, 7 or 8 more stars, and another scoop, and mallet the layers as before; and so repeat. The two layers and the two scoopfuls may fill up, when malleted in, 3/4" in a 9/8 case.

Instead of choking the cases, plaster of Paris is far preferable. Let it be 3/8" or 1" thick, and well dried, before charging. After the cases are charged, bore a hole through the plaster, 1/3 of a diameter, that is, with a 9/8" case, 3/8" diameter. For this purpose it is not necessary to have another shell-bit; bore it with the 3/16", and enlarge it with a penknife. It is better arched under, till conical, as shown in fig. 48. Prime with 4 or 5 pieces of match, and wash with the sash brush.

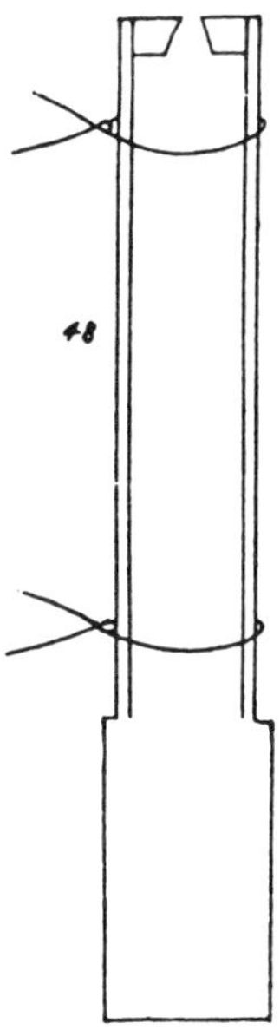

48

The following plan also answers admirably. Saw a cotton-reel in two across the middle; the two pieces will serve for 2 gerbes. Enlarge the hole till it is 1/3 the diameter of the reel, and countersink or underarch it to a funnel shape; glue it in the end of a case of proper size, the flange end outwards, and farther secure it with 2 or 3 screws. Paste a disc of paper over the hole; when dry, set it on a flat surface; drive in starting-fire, so as to fill up the hole, and just cover the reel; then fill up with powder, and stars, as before. Cut away the middle of the disc, to lay the starting-fire bare; surround with a strip of touch-paper; put in a scoop of bursting-powder, and twist to a point. In a large gerbe, plugged with plaster of Paris or with a disc of turned wood, the fire for cascades, may be made to play through a rectangular slit instead of through a circular aperture. The gerbe being finished, make a cylindrical box, or paper bag, of 2 or 3 thicknesses of paper; fill it with a number of crackers and a scoopful of mealpowder, and fasten it to the bottom of the gerbe.

No single piece is more effective than a coloured gerbe; the stars will be projected 20 feet or more: they may be put in mixed; or one layer may be blue, another green, another red. Twist a piece of wire, deprived of its elasticity, round the neck, and another piece round the bottom, and leave long ends; it can then be fastened by them to the top of a post. To remove the elasticity from iron wire, lay it in the fire till red hot; withdraw it with the tongs, and put it aside to cool slowly. If copper wire is used, it will bend without preparation; soft iron wire may also be obtained. String must not be employed, as it might burn, and let the case fall. Common pins, patent short whites, deprived of their elasticity, are useful for connecting the parts of lustres together; put them into a tobacco pipe, and set them in the fire for a few minutes.

FLOWER POTS.

These are choked cases, charged with spur fire: the fire is somewhat slow, so the cases must be short; 4" long, and 5/8" diameter is a good size. Rub the composition thoroughly up in the mortar, preferably with a wooden pestle or bradawl handle; the vegetable black produces beautiful sparks, totally dissimilar to any other, somewhat resembling the rowel of a spur, whence the name. Put a little composition at a time into the case, and jolt it with the Roman candle rammer, not too hard.

Vegetable black, introduced into a nitrate star, causes it to tail, like linseed oil. Light such star on concrete; it will burn, and leave a residue, or cinder, unaltered in shape; blow upon this continuously with the mouth, or, better still, with a pair of bellows: the supply of oxygen will cause it to boil up in a state of fusion, when it will begin to throw out clusters of the peculiar spur-like sparks before mentioned, bright and yellow.

Vegetable black is a pure lamp black; some samples of lamp black make equally good stars, but others are worthless. Greater reliance can be placed upon vegetable black. Vegetable black and lamp black must not be mixed with linseed oil, as such mixture is liable to spontaneous combustion.

Roll up a tube for pill-boxes of two thicknesses of brown paper. When dry, cut it into pieces about 1-1/8" long; choke one end, like a wheel case; set it on a nipple, put in a pinch of starting-fire, then charge it with spur-fire, till full within 1/16": fill up flush with a little plaster of Paris, pressed in flat with a knife: prime the choked end, and put a number of such cases into a rocket head, or shell.

PORT FIRES AND SHELL FUSES.

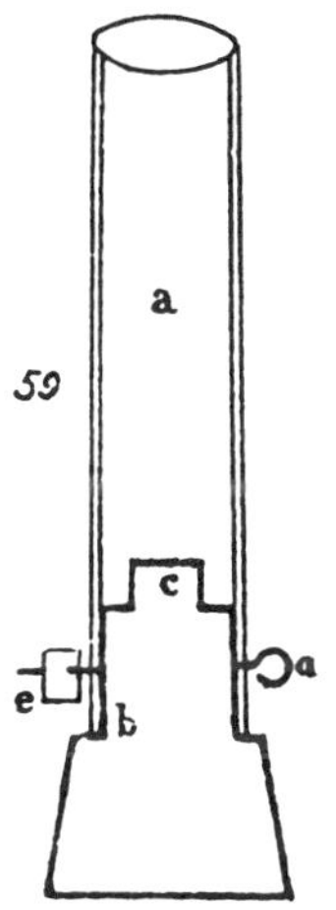

59

262

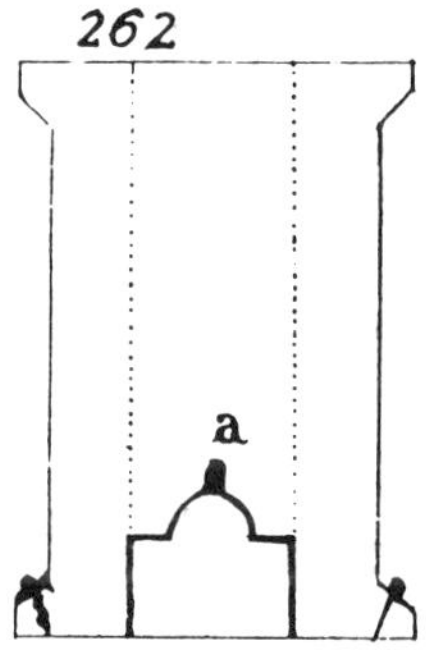

These are unchoked cases, like Roman candles; 6" long, 3/8" internal, 9/16" external, is a good size. They should be rammed, as hard as possible, and for this purpose, it is best to have a mould. Now, if a case is rolled of such a size that it will exactly fit into a brass tube, and is charged in it very hard, it will swell, and it will be almost impossible to get it out again; but if it be made a trifle smaller, so as to just slip through the tube; then, if a piece of writing paper be rolled dry round it once or twice, so as to make it a tight fit, and the case is charged, it can be pushed out, like a pellet from a pop-gun, leaving the writing paper, generally, in the tube or mould, and the case will come out without a wrinkle.

Let the composition be put in very little at a time, and well driven with a solid rammer and mallet. Fig. 59 represents the mould; the foot **b** fits the tube **a**; the tenon **c** fits the case; a wire, **d**, goes through 2 holes in the brass tube, and a hole through the foot; a nut, **e** to keep the wire from jarring out, is made of a piece of india rubber: make a hole through it with a bradawl, and slip it on the wire; or a screw-eye may be passed through and held with a leaden or wooden nut. Of course, rockets and wheel cases can be charged in brass tubes in the same way if desired. The rocket foot will then be like fig. 223.

Small port-fires may be $\frac{1}{4}$" diameter, 3" long, and charged with the funnel and wire. While being filled, they may be supported in a cotton-reel, nailed to a board. Enlarge the hole in the reel, so as to be a very loose fit; its use is simply to keep the cases from falling, not to prevent them from bursting, fig. 262, without the nipple, **a**.

GUM STRIPS.

Cut a sheet of 16-lb. double crown white into 8 equal parts (2.2.2). Take a strip of flannel, 3" wide, 5 or 6 long, and roll it tight into the shape of a cork; secure it from unwinding with a bit of string. Dip the end into a solution of gum arabic, about the consistency of pudding, and rub it over the paper; when dry, cut it into pieces about $\frac{1}{2}$" wide and an $1\frac{1}{2}$" long, something like postage stamps. The best way of preparing the gum is to put 2 ounces into a tumbler of cold water, till the water is about $\frac{1}{2}$" higher than the gum; let it stand 5 or 6 days, stirring it twice daily, till completely dissolved; then strain it through book-muslin, in a funnel supported in a bottle or sample vial.

GUM LABELS.

These are best made of writing paper, thus: suppose they are required to be 1" wide, 2" long; rule the paper in the following manner with ink. Near the edge draw a straight line; at the distance of 1", draw 2 parallel lines, 1/8" apart; at the distance of another inch, 2 more parallel lines, 1/8" apart; and so rule the sheet one way. When dry, rule it at right angles, with a straight line near the edge; at the distance of 2", draw 2 parallel lines, 1/8" apart; at the distance of 2 more inches, 2 more parallel lines, 1/8" apart; and so rule the sheet the other way. Smear the back with gum, as before. When dry, cut between the lines which are 1/8 apart; and the labels will be 2" by 1", with a line on each of the sides for a border, 1/16" from the edge.

TOURBILLIONS.

[Tourbillion, from *Tourbillon,* like postillion, from *postillon,* the *i* being inserted to approximate the pronunciation of the French. In pavilion, from *pavillon,* and vermilion, from *vermillon,* one *l* is dropped; so in battalion, from *bataillon*; while medallion, from *medaillon,* retains the *ll.* For the same reason, the assimilation of sound, *ballon* is changed into balloon, and *marron* into maroon.]

A tourbillion, so called from the French word for whirlwind, is a case made to rotate and ascend at the same time, forming a spiral of fire, and ending in the shape of an umbrella, or hemispherical canopy of sparks.

To Make a Tourbillion.

43

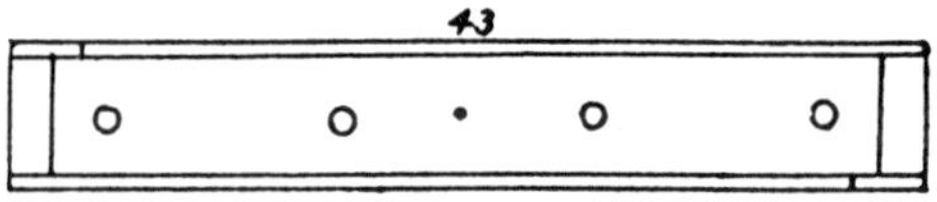

Roll the case like a Roman candle case, but gauge it to the thickness of a rocket case. Let the inner diameter be 6/8"; the outer 9/8"; the length of the case $7\frac{1}{4}$, fig. 43. To charge the case, have a mould, as directed for port-fires; and let the tenon rise exactly 3/8" up the case. Put in a little composition at a time, and mallet it as firmly as possible till within exactly 3/8" of the top of the case; so that there will be a vacancy of 3/8" at each end. Fill each of these ends flush with plaster of Paris. Tourbillion cases must be as accurate as possible, so should the learner find a difficulty in making them precisely, he may roll them a $\frac{1}{4}$" too long; then, when dry, slip one end of a case on a cork, or a piece of wood turned to a similar shape, and cut it rough with a penknife; or fix it on a nose-mandrel in the lathe, and cut it with the chisel; afterwards, the other end. By this means the cases can be adjusted to the greatest nicety.

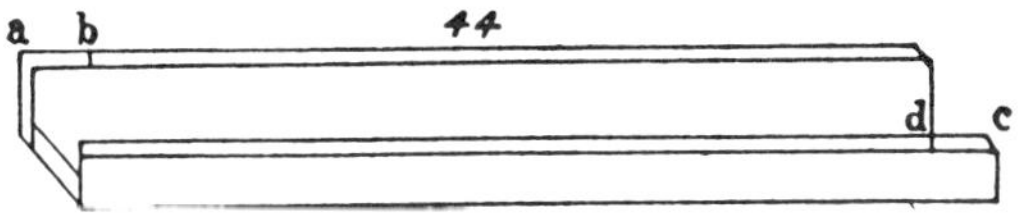

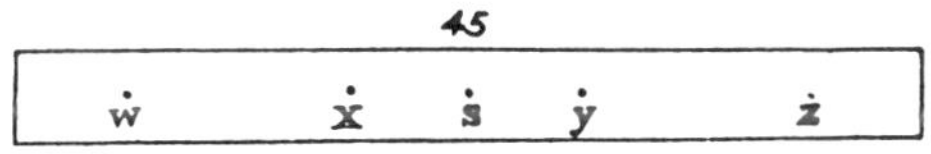

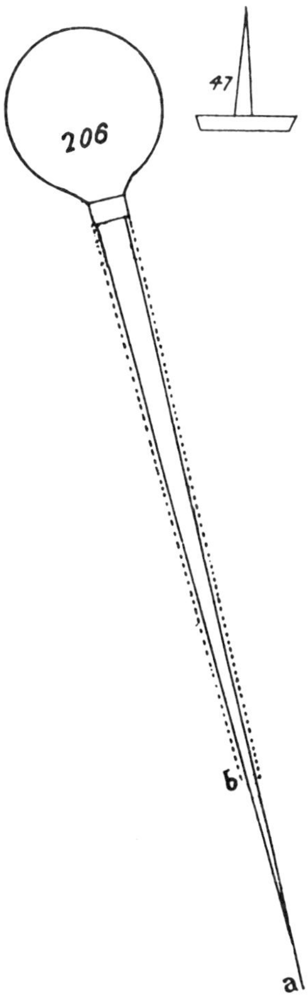

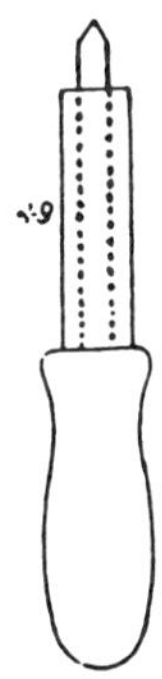

Construct a wooden trough, fig. 44, consisting of a bottom and two sides, firmly screwed together. Each of the pieces of wood is to be 7¼" long and ½" thick. The internal breadth of the trough is to be exactly 9/8"; and its internal depth exactly 9/16", so that when the tourbillion is laid evenly in it and pressed down to the bottom, half of the case will be in it and half out of it. At a point **b**, fig. 44, on the top of the side, ½" from **a**, make an ink mark; and, at a point **d**, ½" from **c**, make another ink mark. Fig. 45 is the bottom of the trough. At a point **w**, 5/8" from the end; and at a point **z**, 5/8" from the other end, make holes with a fine bradawl, truly, in a line down the middle of the wood, as between side and side. The distance **w z** is 6"; divide it into 3 equal parts, in the points **x** and **y**, 2" asunder. Bisect **x y** in the point **s**. Procure 5 carpet pins, fig. 47: they will, probably, be 3/4" long. Drive them through the holes **w, x, s, y, z,** inverting the trough for the purpose, so that they shall stand bolt upright in the trough. Now screw or nail a piece of wood over the bottom of the trough entirely to cover it, to prevent the carpet pins from getting displaced. It will be seen that if the tourbillion be now laid evenly in the trough, and pressed down till it rests on the bottom, the projecting pins will make 5 holes, in the under part of the case. While it is thus lying, with a stiletto, such as used by seamstresses for making eyelet holes, fig. 206, prick the side of the case over the line **b** of fig. 44; and also over the point **d**. There will now be 7 holes; 5 underneath, 1 to the right and 1 to the left: the latter are the places for the whirlers, or holes of rotation; 4 underneath, for the risers, or holes of ascension; the centre one, **s**, receives a nail or screw. Take a bradawl, fig. 49, 3/16" diameter; and slip over it a shield, consisting of a piece of wood with a central hole up it, like a pop-gun, of such a length, that when it is slipped on, only ¼" of the bradawl protrudes; or instead of a bradawl, fix in a handle a brass wire of equal length projecting, namely ¼", and file it to a conical point. Push this into all the holes, except **s**, making 4 holes underneath, and 2 horizontal holes, one left, one right; all these holes will be exactly of the same depth on account of the shield; see that they are bored perfectly true, the horizontals exactly 90 degrees above the others, or ¼" of the circumference.

The next thing required is a piece of hooping, or curved stick, fig. 46, about 6/8" wide and as long as the case, 7¼". In the centre of this, bore a hole, and countersink it; then, with a screw, 1¼" long, screw the hooping at right angles, on the bottom of the case, through the point **s** of fig. 45, which must be enlarged to receive

214

the screw; or the hooping may be fastened with a clout nail, fig. 214. A touch of glue may still farther hold the wood in position. It will now assume the shape of a cross, like figs. 50 and 51. Fig. 50 shows the under side of the case; fig. 51 the upper. From **w** to **x** lead a bit of naked match; push the ends into both holes, and secure with a little wetted meal pressed in with a knife. Do the same with **y** and **z**. Paste a piece of tissue paper, 3/4" wide, and of sufficient length, and cover each of the two pieces of match with two layers of the paper. Turn it over, like to fig. 51; connect the holes **a** and **b** with a bit of naked match; and under the centre of it slip another piece of naked match, having a piece of touch-paper round the protruding end; cover the match with two thicknesses of pasted paper, in the same manner as the under holes. The tourbillion is now complete. See that it will balance, and swing round easily when laid upon a level surface. The proper way to fire it, is from a flat sheet of iron, or a flagstone. Light the touch-paper; the fire will communicate to the side holes, and set it in rotation. As soon as 1/8" of powder has burnt from each end, and the piece has got well into action, the 4 under holes will catch, and cause it to ascend.

50

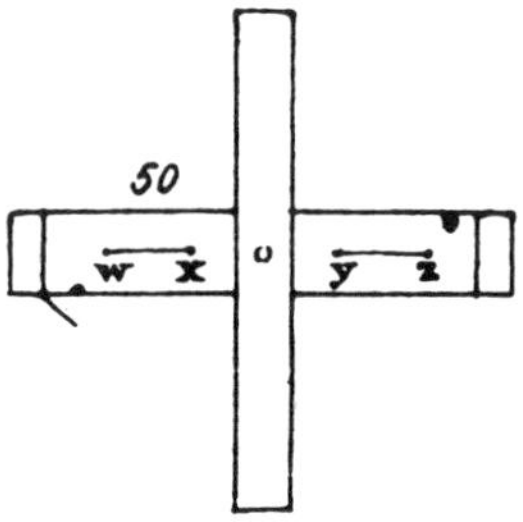

51

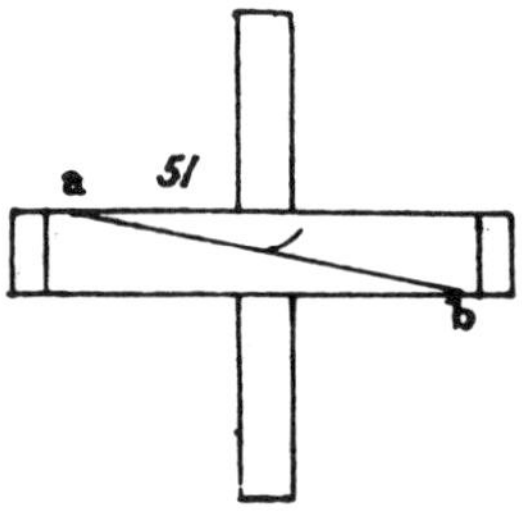

Instead of this mode of making a tourbillion, some charge it with an inch of solidly-rammed clay in the middle; fasten the stick by crossing it with binding wire; bore a hole through the middle of it and of the clay, and slip it over a tapering-wire, standing upright in a block like the spindle of a rocket. Four holes only are then used; two of rotation and two of ascension; and the whole are fired at once, the match starting from one of the under holes, going to the side hole; over, across, to the other side hole, and on to the other under hole.

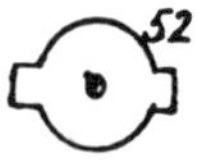
52

Instead of making them with clay in the middle, there is yet a better method of having two cases, each about 4" long, and gluing, or fastening them with tin-tacks, on a centre-piece, turned with a tenon at each end, fig. 52; two balancing arms, one on each side, must then be fixed to the centre-piece.

In a windmill, as is well known, the vanes, or sails, are set at an angle. There is a toy, made of two slips of tin forming a cross, and set at an angle, sloping upwards, called the Flying Dutchman; this, when spun with a string from a handle like a humming-top, flies up into the air on escaping from the string. Steel Fliers, with two vanes, are used by sportsmen to practise shooting flying. Small balloons,

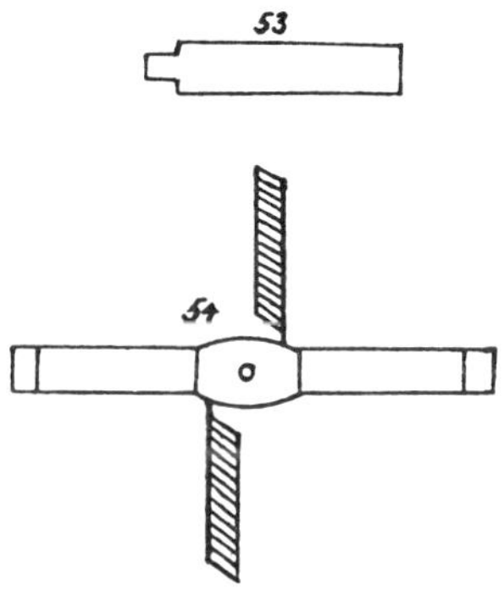

some years ago, in a room in the Polytechnic Institution, free from a current of air, were guided or driven by a similar contrivance, moved by clockwork. The screw propeller of a ship acts on the same principle. I think it possible that, if two vanes were fixed in the central piece of wood, set at an upward angle of 10, 15 or 20 degrees from the horizontal, they might assist the ascension, and so cause the tourbillion to reach a greater height; or, the vanes alone might cause it to rise, upon 4 side holes, two to the left and two to the right, causing it to rotate. The design is shown at fig. 54; the shape of the vane at fig. 53. I offer it only as a suggestion.

Winged Tourbillions.

Tourbillions, made in the manner described earlier, are apt, at times, to take an erratic flight, or to pitch over, and lie kicking about upon the ground. To obviate this defect, contrivances have been resorted to, called cradles: frames made to turn upon a pivot, set in motion by the tourbillion itself and supposed to retain it in equilibrium, until it can commence its ascent. The tourbillion, however, on quitting a thing in motion, is like a person making the same attempt, and is more apt to be thrown to the ground than to rise into the air.

It struck me, during the sale of the first edition of this work, that the failure must be owing to some defect inherent in the construction of the tourbillion itself. Everyone knows the danger of persons in a panic rising suddenly in a boat, and thereby giving it a rocking motion, and throwing the centre of gravity over the edge. If a wagon be loaded so high, that when the wheel sinks into a rut, the line of direction, from the wagon's centre of gravity to the earth's centre of gravity falls without the wheel, the vehicle will be upset.

In a skyrocket, adjusted to rise properly, the centre of gravity is in the stick, just below the mouth of the rocket; and as the powder, burning, lightens the head, it shifts lower and lower, and keeps the stick downwards, and the rocket erect. If the rocket were fitted with too short a stick, the centre of gravity would be thrown above the choke, and the rocket would pitch over. Now, in nailing the stick *under* the tourbillion, the centre of gravity is raised *above* the level of the balancing power; so that it would seem as if it were designedly rendered top-heavy, and constructed purposely to overturn. Where are the wings of a bird, or butterfly? *Under* the body? Certainly not. The moment the idea of this comparison occurred to

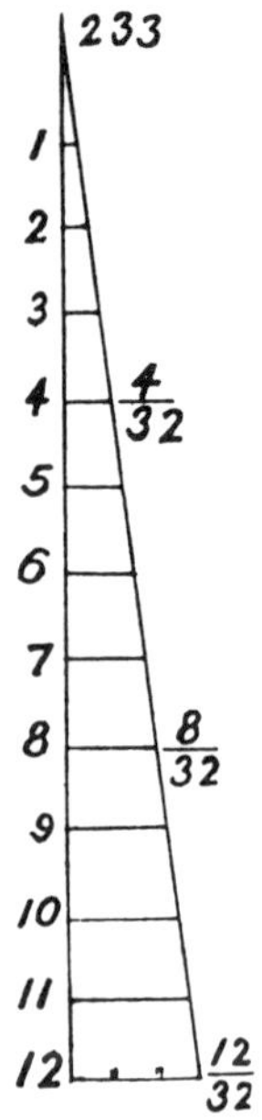

me, I determined to take nature for my guide, to copy the bird, to imitate the butterfly. I nailed the stick, the equipoise, the outstretched *wings* upon the *top* of the case; and fastened a pin-wheel block underneath, as a foot, for it to stand and turn upon. The immediate consequence was perfect success; the tourbillion spun round, and rose perpendicularly. I have made scores since of all sizes, down to so small as 2/8", without a single failure. The learner can study the following table and experiment for himself. The diameter of the holes is shown for each size at fig. 233; but the same hole will answer for two sizes.

Bore	Distances in eighths of an inch	Total	Length	Holes
3/8	2,2,1,10,3,2,2,3,10,1,2,2	40	5 inches	1/8 inch
4/8	2,2,1,12,3,3,3,3,12,1,2,2	46	5¾ inches	1/8 inch
5/8	2,2,1,14,4,3,3,4,14,1,2,2	52	6½ inches	3/16 inch
6/8	3,3,1,15,4,4,4,4,15,1,3,3	60	7½ inches	3/16 inch
7/8	3,3,1,17,5,4,4,5,17,1,3,3	66	8¼ inches	¼ inch
8/8	3,3,1,19,5,5,5,5,19,1,3,3	72	9 inches	¼ inch
9/8	4,4,1,20,5,5,5,5,20,1,4,4	78	9¾ inches	5/16 inch
10/8	4,4,1,22,6,5,5,6,22,1,4,4	84	10½ inches	5/16 inch
11/8	4,4,1,24,6,6,6,6,24,1,4,4	90	11¼ inches	3/8 inch
12/8	4,4,1,26,7,6,6,7,26,1,4,4	96	12 inches	3/8 inch

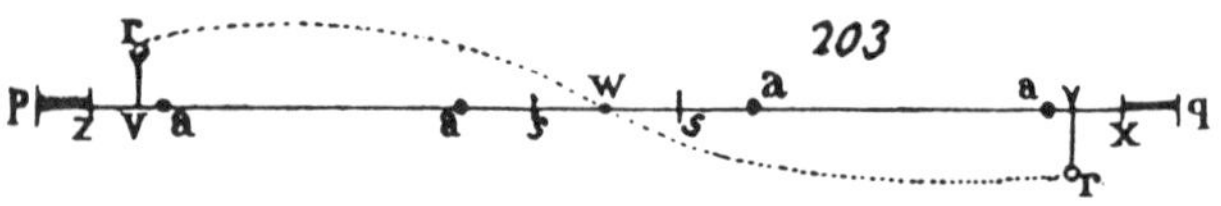

Fig. 203 represents the distances of a tourbillion, say a 4/8", laid down to a diminished scale. From **p** to **z** is 2, the plaster of Paris; from **z** to **v** is 2, the distance for the first whirler **r**; from **v** to **a**, 1, to allow the piece to get into rotation, before firing the riser, **a**; from **a** to **a** is 12; from **a** to **s** is 3; from **s** to **w**, 3, the centre; the distances are then reversed. The distance from **s** to **s** is the breadth of the stick; from **p**to **q**, is the length of the stick, which is to be the same as the case. Taper the stick towards each end, to imitate a wing, and to offer as little resistance to the air as possible.

To Make a 4/8 Winged Tourbillion.

The length of the case, as shown by the table, is 5-3/4". Cut a long strip of kraft paper, 5-3/4" wide; paste it, and roll it on a 4/8" former, till the outside gauges 6/8", like a corresponding rocket. The first number in the table is 2, which shows that the plaster is to be 2/8" thick: fit the case on a foot, fig. 9, turned with a tenon to enter 2/8"; drive in tourbillion powder till full within 2/8"; remove the foot, and fill each end of the case with plaster of Paris. Procure

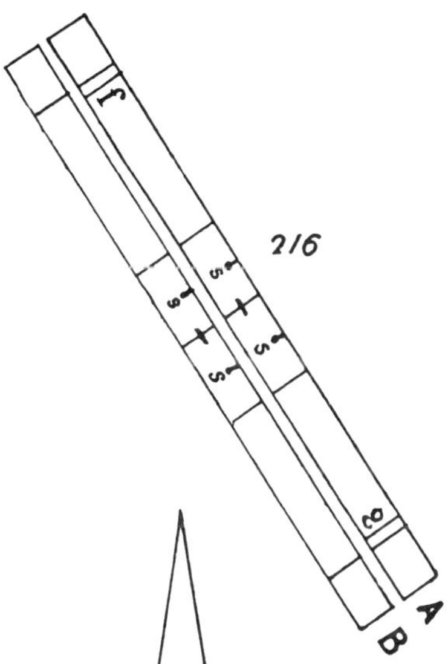

216

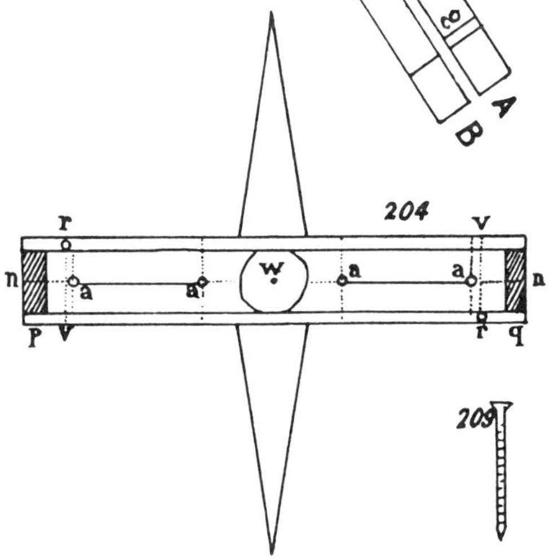

204

209

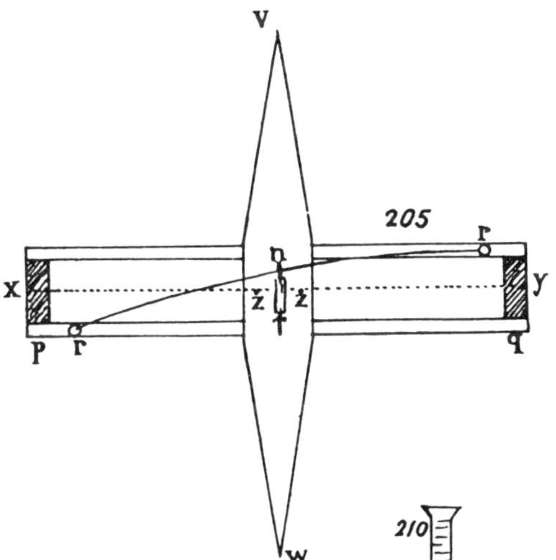

205

210

a piece of pine, about 1/8" thick; a venetian-blind lath is very convenient, and may be obtained of the makers, before being bored, or painted; if not, plane a piece of pine to that thickness. Cut from it a strip as long as the case, and as wide as the middle two numbers added together, that is, 5-3/4" long, and 6/8" wide, shown on a diminished scale at A, fig. 216. Adding together the first two, and the last two, distances given in the table, they become 4. 1. 12. 3. 3. 3. 3. 12. 1. 4: set these off upon the strip and write upon the back of it, "4/8 tourbillion gauge". Cut another strip of the same length and breadth, and taper the ends to points, to imitate a wing, as shown on a smaller scale at **vw**, fig. 205. Draw a straight line along the case, **xy**; lay the gauge by the side of it, and mark off the 2 distances, **ss**, fig. 216, for the position of the wing. Find, by the same gauge, the middle of the wing; draw a line across it; lay it over the middle of the case, crossing it at right angles, between the 2 marks just made; pierce 2 small bradawl holes, **zz**, fig. 205, and drive in 2 rivets, fig. 209, or 2 tintacks. Now turn the tourbillion over, and lay it flat upon the wing, case upwards; draw a straight line along the middle, **mn**, fig. 204; upon this set off all the distances from the gauge, fig. 216, except **ss**, which have been already set off on the other side. At the first distance, that is, $\frac{1}{2}$" from the end of the case, wind a piece of cardboard round it, and pencil a mark by its side; draw a straight line across the plaster at each end, connecting the points **m** and **n**, fig. 204, with the points **x** and **y**, fig. 205; cross each of these lines, on the plaster, with another passing through the centre at right angles; and, from one of its extremities, draw a short line along the case, parallel to **mn** and **xy**; and, of course, half-way between them, till it crosses the line drawn round the case; this will give the position of the whirler, **r**; the same with the other end.

Now procure a pin-wheel block; enlarge and slightly countersink the central hole; make a bradawl hole at **w**, fig. 204, and nail the pinwheel-block to the case. The best kind of nails for this purpose are stubs; they are of the shape and thickness of fig. 210, about 5/8" long, with a sharp point and a flat head; they are a species of rivet, and look very much like screws, only that the ridges round them are collateral indentations, not a continuous spiral, and the heads have no slit in them; they are sold at the grindery shops, at 5*d.* per lb., which contains about 600. Should there be any difficulty in obtaining them, flat-headed screws will serve. Now make the holes **aaaa**, fig. 204, with a bradawl, 1/8" diameter, shielded, fig.

49, so as to enter the case ¼"; likewise the whirlers **r** and **r**. For large tourbillions, 8/8" and upwards, the holes may be made with a punch; work the punch round till it just cuts through the case and lays the powder bare. Connect the risers **aa** on one side of the centre **w**, with a piece of match; then the two on the other side; and cover them with a strip of pasted tissue paper. Turn the tourbillion over, and set it on the pinwheel-block foot; the centre of gravity being now below the level of the wings, the piece has no inducement to overturn; it may be swung round with the greatest rapidity, without upsetting, and will show what may be expected from it. Cut a bit of match, about 1½", and a piece of touchpaper 1½" square; lay the match on the touchpaper, so that it lies half in and half out; put with it a pinch of mealpowder; roll the paper up, and twist to a point. Cut another piece of match, about as long as the case; push one end of it into the left hole **r**, fig. 205; lay the naked end of the touchpapered piece of match along the stick, as at **nt**, between the rivet heads **zz**; pull the other match over it, and push the end into the right hole **r**. Cover with 3 pieces of pasted tissue paper; one on each side, and one over the middle. A piece of sheet iron, slate, or board, 7" or 8" square, will be large enough to fire it from, as it will spin round, without wandering, if carefully made. Pay particular attention that every hole is bored of the same depth, and that one half of the tourbillion is the exact counterpart of the other. If the head of the stub be not quite buried in the countersink hole, but allowed to protrude very slightly, on the thickness of a playing-card, it forms a pivot, which permits the tourbillion to rotate with still great freedom.

Tourbillions may be made with only 4 holes, 2 risers, and 2 whirlers, all fired at once, as in fig. 219. The gauge for marking these is the same as for the other, simply omitting the lines **f** and **g**, on **A**, fig. 216, and leaving it like **B**. The order of fitting up is always the same: first wing; then foot; then leaders, **W**, **F**, **L**.

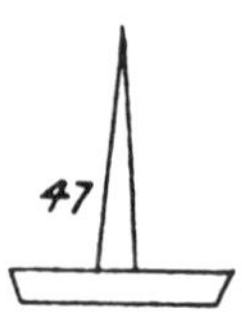

Carpet-pins, fig. 47, having turned heads, make excellent feet, superior to stubs, but they are dearer; have a pin-wheel block, or thin disc of cork between, to clear the leaders.

COLOURED TOURBILLIONS.

These must not be smaller than 6/8". Having charged a case, as fig. 217, before putting in the plaster make 2 bradawl holes at each

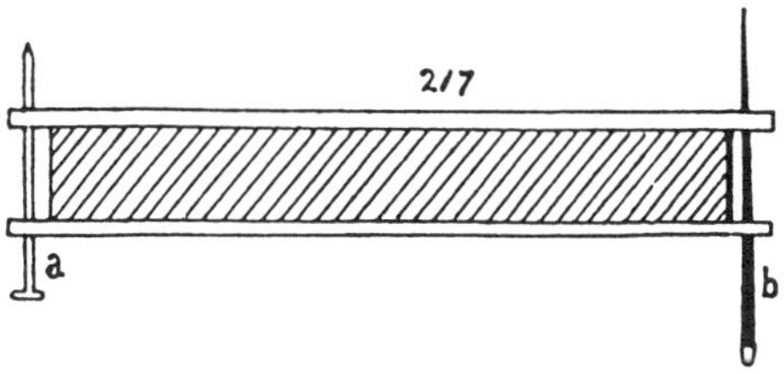

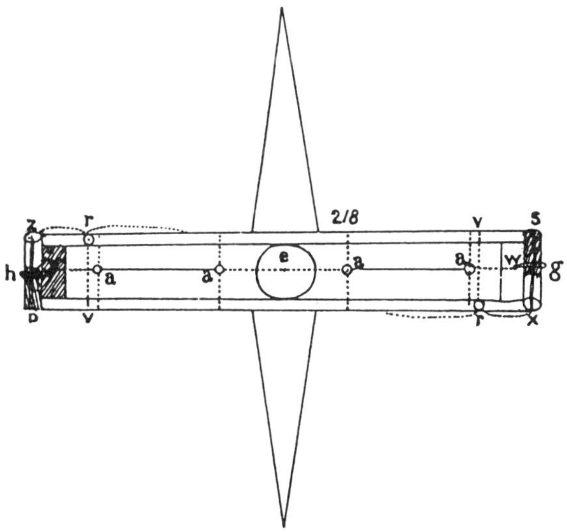

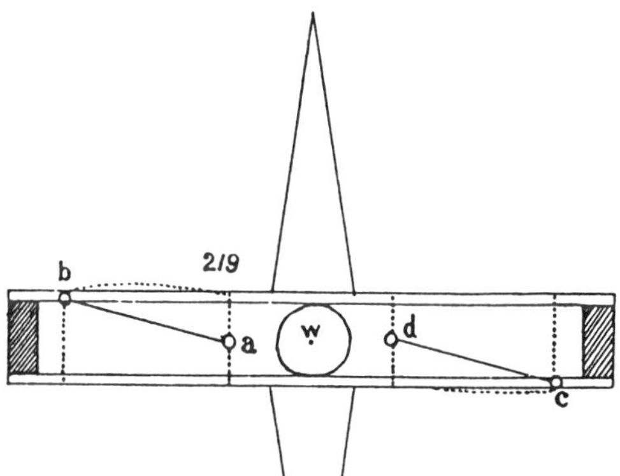

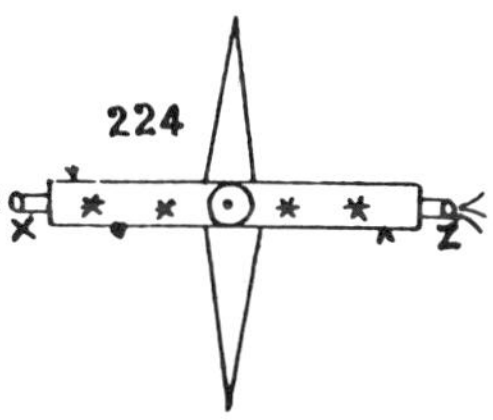

end, and push through them a greased, or oiled needle, or French nail, as **a** and **b**. Fill the ends with plaster; when dry, pull out the nail, or needle, with a pair of pincers, a hole will be left to receive a wire. Roll up a colour case as long as the external diameter of the tourbillion; thus, for a 6/8", cut a piece of writing paper, 4½" by 2¼"; paste it all over, and roll it on a 7/16 former, so as to make it 2¼" long; when dry, cut it across the middle, to produce two cases, each 1-1/8" long. Set one on a flat surface, put in a pinch of meal-powder, then some coloured fire, and drive it in, till it fills up half an inch; fill the void at the end, **hp**, fig. 218, with fine pine sawdust, pressed in tight, till nearly full; then stop up with plaster. Through the hole **v**, fig. 218, left by the nail **a**, fig. 217, insert a piece of thin binding wire **hv**, and twist it round the colour; the same with **wg** round the colour at the other end. Care must be taken that the holes **v, a, a, e, a, a, w,** all lie in a straight line; and the mouth of the colour at **z** must point in the same direction as the adjacent whirler **r**; and the mouth of the colour at **x**, in the same direction as the other whirler **r**. Connect the risers **aa, aa,** in the usual way; run the other leader from **z** to **r**,and thence, over the wing, from **r** to **x**; the match may be tied in position with a piece of thread or hemp, **zp, sx**. Envelop the whole, on each side, with a piece of pasted tissue paper. It is essential that the colours be fastened on sidewise; were they pointed endwise, as at **x** and **z**, fig. 224, the centrifugal motion is so violent, that the composition would be instantly whizzed out in flakes. By fixing them, as in fig. 218, the tourbillion will carry a beautiful ascending ring of coloured fire. Sugar blue No. 1; blue, 6, 13, 17; red, 1, 6, 7; green, 7, 8, 9; and the magnesium colours are recommended: they may be damped. Instead of wings, a strip of tin, of the shape and size of the gauge, fig. 216, may be nailed on the top of the tourbillion, and the ends twisted up to an angle, like a screw-propeller, or "the Flying Dutchman", as suggested. I have since tried it, and it will answer: it will also raise the tourbillion, by the mere rotation of the side holes, but not to a sufficient height to compensate for the extra trouble. It is not worth trying, except as an experiment.

To Make a Spile Rod.

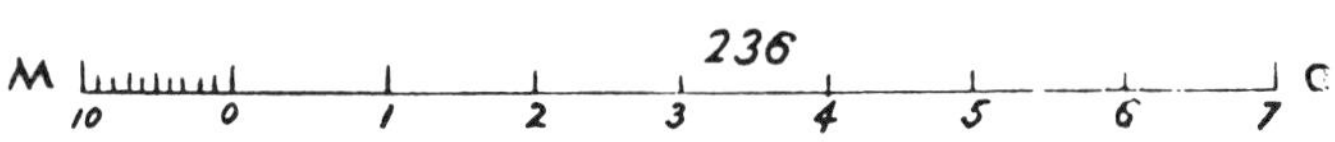

Cut off, and plane smooth, a piece of venetian blind lath, 1/8" square, and 14" long, cover it with a piece of pasted tissue paper, 14" long, and 1-5/8" wide. When dry, divide one of the sides into inches and eighths, and another into inches and tenths, as shown

187

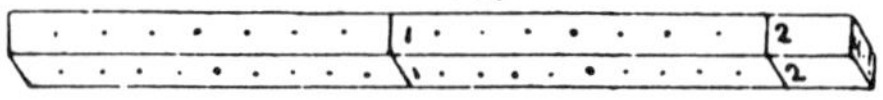

at fig. 187, which is drawn of the correct size, except as to length. A third side may be divided into centimetres and millimetres, which are given at fig. 236: the centimetres, **C**, to the right; the millimetres, **M**, to the left. This instrument will be found convenient for taking the internal length of partly filled cases, depths of vials, etc.

SAXONS.

55

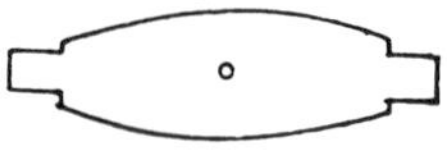

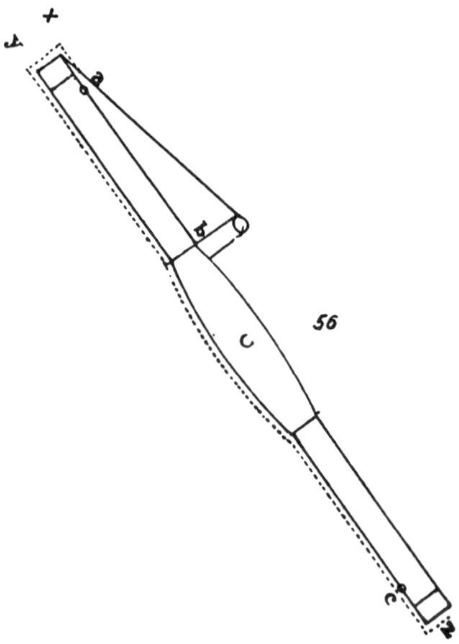

56

These are unchoked cases, charged like a tourbillion, but pierced only with holes of rotation, for the purpose of turning a coloured fire. Drive them in a mould, as directed before. A good size is, 5/8" internal, 7/8" external, 6" or 7" long. Let the tenon enter the case 3/8": charge the composition firm till within ½" of the top, which leave vacant. Remove it, and fill the 3/8" occupied by the tenon, with plaster of Paris. Have a centre-piece, turned like fig. 55, with a tenon, 5/8" diameter at each end, ½" long. Glue a case on each tenon. Let the centre-piece be 6" long, exclusive of the tenons; so, if the saxon cases are 7" long, each, the entire length, as fig. 56, will be 20". Make a hole at **a**, and another at **c**, with a shielded bradawl, 3/16" diameter. Put a bit of naked match in the hole **a**, carry it round **x** and **y**, along to **c** and on to **z**. It must be pushed into **c** with a blunt wire. Cover it with two thicknesses of pasted paper, like the tourbillion. Leave the match exposed at **x**, **y**,, and at **z**, and brush it over with meal paste. If the central piece of wood were now put on a horizontal spindle, and fire communicated to the match **z**, the holes **a** and **c** would cause it to rotate, and produce a white circle of fire. This, however, would be hardly worth making; but, by fixing at **b**, a little case of coloured fire, a splendid effect is produced. This case of colour is usually tied to a nail, driven in at **b**; but a little tenon of wood may be glued there instead, and the case of colour must be then charged, with a vacancy at the bottom, to fit on the tenon. The case of colour must be timed to burn as long as the saxon; rather more than an inch will be sufficient.

Saxons are sometimes made by charging a Roman candle case with 1" of clay in the middle, and boring a hole through the clay, to receive a spindle. Only one half of the case burns at a time; a leader, placed at the bottom, near the central clay, conveys the fire to the other end; and continues the rotation. For distinction, they are called Chinese fliers.

FIVE-POINTED STARS.

These are unchoked cases about 2½" long, and 1" diameter. Make a bottom to the case with ¼" thickness of plaster of Paris. Charge it solid, and at 3/8" from the extremity, that is, 1/8" beyond the plaster bottom, round the circumference, make five holes, as for saxons; run a bit of match round, connecting the holes. When fired, they stand out at right angles, the plaster towards the spectator, so that the fire resembles a gas star with 5 points, as in fig. 130.

SQUIBS.

Take a sheet of 60-lb kraft, and a sheet of 12-lb. white demy. The kraft, as said before, is 29" by 22½". Cut it into 24 equal parts (2, 4, 3), that is, first into two equal parts, down the natural fold of the paper; then each into four equal parts, at right angles to the first fold; and each of these into three equal parts, at right angles to the second fold (p, h, p, al, perpendicular, horizontal, perpendicular, alternate). Each piece will then be 5-5/8" by 4-5/6". The demy is 22½" by 17½". Cut the sheet into sixteen equal parts (2, 4, 2), each piece will be 5-5/8" by 4-3/8". A piece of kraft and a piece of white will make a case 5-5/8" long. For a former, procure a piece of brass wire, or stair-rod, about a foot long, and ¼" diameter. Lay eight pieces of the demy evenly on each other; draw the thumbnail of the right hand from the farther edge of the paper straight over the middle towards you, a few times. If properly performed, this will draw piece behind piece; proceed till about 1/8" of each is left exposed, in the same manner as a pack of cards would arrange themselves, if set upright, and allowed to fall forwards; something after the manner of the laths of a venetian blind, or slates upon a roof. If you cannot acquire the knack of doing this, you must so lay them, one by one. Paste or gum the edges of all the eight pieces thus lying. Place one of the pieces of brown paper before you; lay the former, or stair-rod, across it, nearly in the middle as, **a b**, fig. 253; bend the paper over it, and press it in with the fingers of both hands; roll for an inch or so; lay it on the middle of a white piece; bend the white over; infold the brown in it, and roll forward, till it catches the pasted or gummed part, and sticks. After a number have been prepared, and are dry, choke them, as directed for rockets; the paper will be sufficiently soft, not having been pasted. The wire of the choker, fig. 24, should be about 1/16" thick. Now take a piece of square steel, or iron wire, 12" long, the thickness of the wire being about 3/5 the diameter of the stair-rod, that is, ¼ x 3/5 = 3/20"; if it be found difficult to procure square wire, file a round

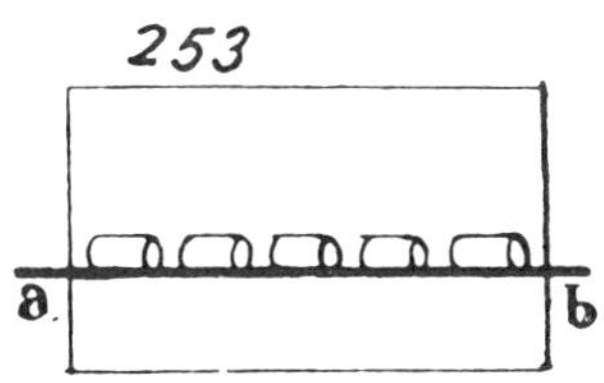

65

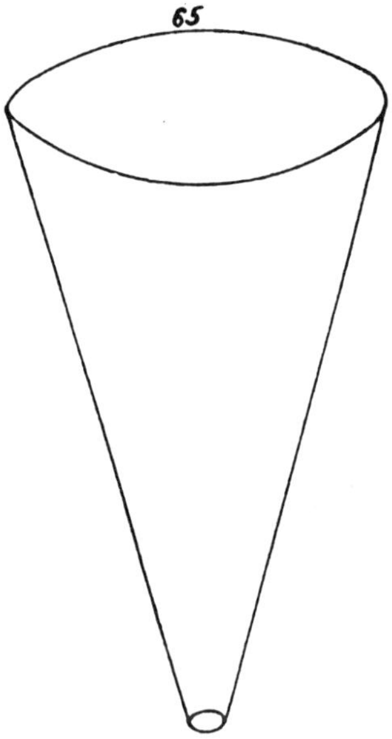

4

piece. Bore a hole down a bradawl or chisel handle, 2" in depth, and $\frac{1}{4}$"; fix the square wire in it, with melted lead, like fig. 4; the lead is for the purpose of giving weight to the blows in charging. Black the wire all over with ink, and alloy to dry. The next requisite is a tin funnel, without a neck, of the size and shape of fig. 65. Any tinman will readily make one to order; but if the learner procure a soldering-tool, or copper-bit, he can construct one himself. For this purpose, dissolve a piece of zinc in a little hydrochloric or muriatic acid, till the acid is "killed", that is, saturated: heat the tool, and dip the tip end, momently, into it; the acid combines with the oxide of copper formed by heating, and the zinc adheres to the clear surface of copper produced; it will now easily take up the solder: the joint to be soldered must be clean, and also touched with a feather dipped into the acid. It is best to make a funnel with a piece of writing-paper, first, for a pattern; when this is got correct, the tin can be cut according to it. By describing a circle with a 3" radius and cutting out a sector of 100 degrees, the correct size is obtained at once; 1/8" breadth, outside the radius, is to be allowed to lap over. The bore of the squib is 5/20"; the size of the ramming wire 3/20; the hole at the bottom of the funnel may be half-way between, 4/20 or 1/5" in diameter. Push the wire rammer through the funnel, till it protrudes 2" beyond the bottom; observe the part of the wire which is now level with the top of the funnel; withdraw it, and file a bright mark round the part: it will be about $4\frac{1}{2}$" from the end. The wire, having been inked, shows the bright mark more plainly. The square steel wires of umbrellas may be used; but, as they are scarcely thick enough, file the end bright, dip it into the acid and coat it with a little knob of solder: wires so prepared let the composition run down easily, and readily consolidate it.

2/2

Make a pine box, 3" square and 4" deep, and nail a bottom to it. Also have a little nipple, fig. 73, with a wire, as drawn; it can be secured to a flat board; a hobnail, fig. 212, also makes a good foot.

To charge the cases. Set the choked end of one on the nipple or hobnail, insert the funnel in the other end, put the wire rammer down through the funnel, and let it fall to the bottom of the case: then put in some powder, and jolt the rammer up and down, till the case gets so far filled, that the bright, filed notch, before described, is on a level with the top of the funnel. It is necessary to sit at a table to charge the cases; the eye is, then, on a proper level to see the mark. As the cases are thus charged, set them in the square box, choked end downwards, till a number are filled. Now take out a handful, invert them over a sheet of paper, and give them a few taps with the rammer; this will shake out a great deal of loose composition, that has got puffed up, in the case. Return them to the pine box.

The next operation is to bounce, or bang them. For this purpose take a long slip of paper, 7" wide. Stretch it lengthwise before you. Lay a number of the cases upon it, so that all the choke ends lie flush with the left edge of the paper; then roll them up in the paper; turn and set them upright on the choke ends; the paper will now stand up more than an inch above the cases, as a tumbler, three parts full of water, stands up with an empty space above the water. Rest them on a sheet of paper, and pour in a quantity of F grain powder; this will fill every case; loosen the paper wrapper, and alloy the superfluous grain powder to fall on the paper. Set the cases, choke downwards, in the square box. As they are all full, it is necessary to get a little out of each. For this purpose, lay the square box, containing them, horizontally on a sheet of paper, on one of its sides; turn it gently over, and lay it on the next side, raising it a trifle to make it slightly slope; a little of the gunpowder will spill out; turn it gently over again, and lay it on the next side; and so proceed, till every case has about 3/8" empty; then set the box upright. The ends of the cases are now to be closed. Take a case in the left hand, as in fig. 63, and wind a piece of string, or whipcord, three or four times round it, holding the end of the string, **a**, firmly with the thumb; then, with the right hand, bring the end, **b**, back over all the folds, as in fig. 64; hook the left forefinger round the end, to prevent the string slipping; pull the end, **B**, tight, and the case will be closed.

63

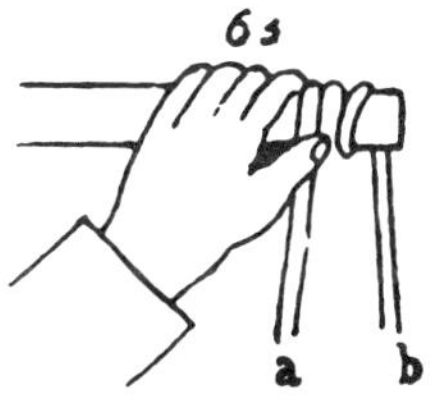

64

Melt some paraffin wax in a double boiler; carry it to a distance from the fire, and dip the ends of the squibs into it. Next, prime them, by pressing the choked ends into very slightly damped meal powder, as directed for wheel cases. Take a piece of touch-paper, 15" by 10". Divide it into 48 pieces (2, 2, 3, 4); each piece will be $2\frac{1}{2}$" by $1\frac{1}{4}$". Roll a piece round the primed end, twist to a point, and fasten it with a bit of carpet thread, hitched three times round it.

SERPENTS.

These are simply squibs, made short, in order to burn out quickly, as they are intended to be thrown from mines, and would other-

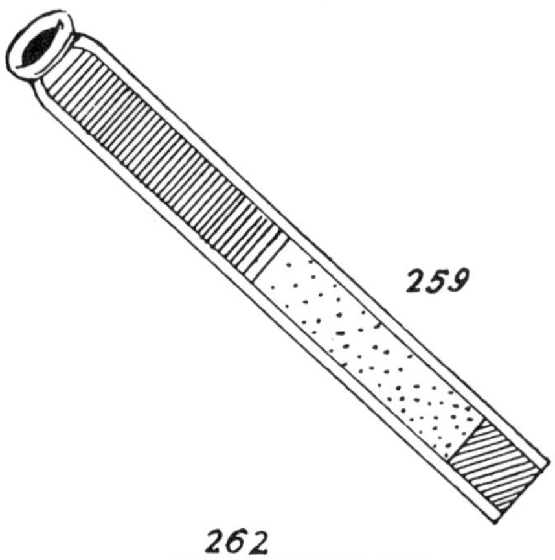

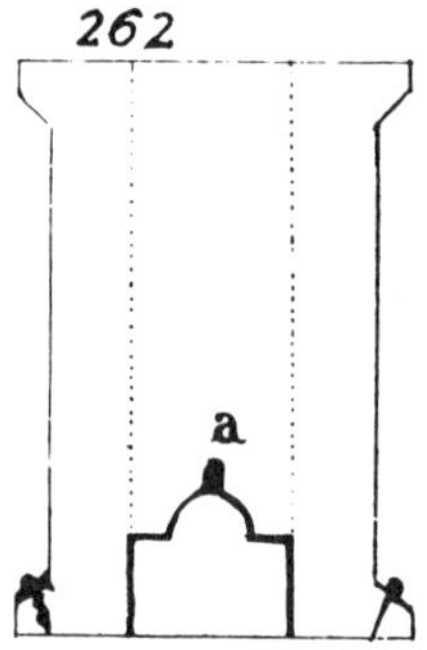

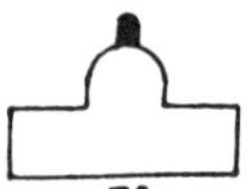

73

wise lie too long on the ground. Cut the 60 lb. kraft paper into 48 equal parts (2, 4, 3, 2), each piece will then be 2-13/16" by 4-5/6". Cut the demy into 32 equal parts (2, 4, 4), each piece will be 2-13/16" by 4-3/8". A piece of kraft and a piece of white will make a case 2-13/16" long. Roll the cases dry on the squib former, and choke them in the usual manner. Ram them with the funnel and wire, but continue the ramming till the guide-mark stands a little higher than the level of the funnel; until, in fact, the cases get half full. Bounce, close, and dip them in melted wax, like squibs; or plug them with plaster of Paris, which method makes equally good reports, see fig. 259. The best way to prime them is to paint them with the sash brush, slightly, then plunge them into dry meal, so that they may be, as nearly as possible, like quickmatch. If they were pressed into wet meal powder, of the consistence of mortar, and then rubbed smooth on a Dutch tile, or slate, or plate, they would dry hard and glazed. In this case, if they were fired from a mine, in all probability three fourths of them would miss. In brushing them with the sash brush, with meal paste, turn the brush well round, to leave little in; so that, after being pressed into the dry meal, they may not be choked up, but the cup shape still be left. They are not to be touch-papered, being intended only for mines, rockets, or shells. They may be supported in a cotton-reel nailed to a board while being charged; see fig. 262. It will be necessary to enlarge the hole in the reel with a nose bit or twist-drill, and to fit it with a nipple, like fig. 73.

TOURBILLION SERPENTS.

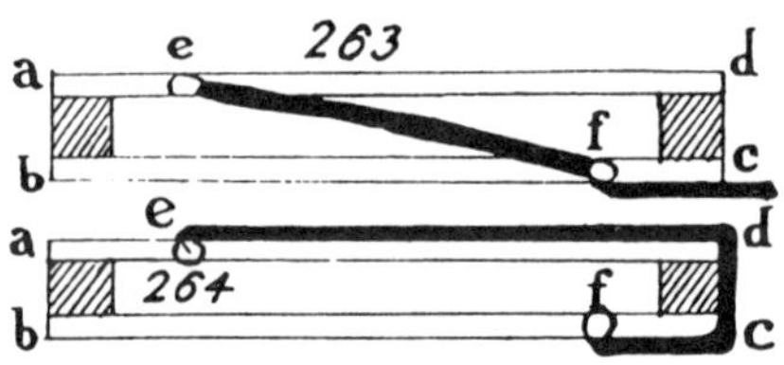

Roll a case 3" long on the squib former, with brown paper, pasted all over. Charge it with tourbillion powder, and stop each end with plaster of Paris. Run the match, fig. 263, from **e** to **f** and on to **c**; or from **e** round **d** and **c** to **f**, fig. 264. Have a piece of thin tissue paper, as long as the case; paste it, and roll it round, so as to leave the match exposed at **c**, fig. 263; or the end **d c** at fig. 264. These may be put into rocket heads, mines, or shells.

PINWHEELS.

Procure some 16-lb. tissue paper white paper. A sheet is 30 x 20". Cut off a strip 30 x 3", which leaves a piece 30 x 17".

The small strip may be used for odd purposes. Divide the 30" by 17" into 10 strips, each 3" by 17", for penny pinwheels. If the paper were cut the other way of the sheet, it would not wind smoothly, but crumple up. The fibres, from some cause or other, appear to arrange themselves in one direction, like the grain in wood. For halfpenny wheels, divide the sheet into two pieces, each 10" by 30"; then cut each of these into 12, each 10" by 2½". For a former, have a straight piece of iron, or steel wire, 1/8" thick, and 24" long. One end of this must have a basil, or sloping enlargement upon it, which is thus made: Take a piece of tissue paper 4" square; lay it straight before you and cut it diagonally, from the right top corner to the left bottom corner; remove the left half; paste the right; lay the wire upon it so that the rectangular corner is 5" from the right end of the wire; roll it up and press it smooth; the sloping edge will now form a spiral, commencing at 9" from the right end of the wire and being enlarged at 5" from the end. Paste a thin strip of paper over the whole, to protect the spiral edges. It will now be something of the shape of fig. 27, only turned the reverse way. The 5" at the end are for a handle. To roll the pinwheel pipes, lay the strips of paper evenly upon each other, and work them back with the thumbnail, as before. Paste the edges; lay half-a-quire of tissue paper, or of blotting-paper, flat upon the table, to roll on. Place one of the strips of pasted paper upon it; lay the wire upon it, the basil end being towards the right hand; the wire is to be laid, not quite parallel with the near edge of the paper, but slightly sloping, about 1¼" distant from it on the left, and not quite an inch on the right, or basil end. So much of the basil must lie on the paper as will make the mouth of the pipe large enough for the nozzle of the funnel to enter ½". The basil, of course, gathers up the paper more quickly than the wire, and brings the edge straight, on the completion of the pipe. Now to roll, bend the paper over the left end of the wire, press it in with the fingers, and begin immediately to move it forward; at the same time draw the fingers of the right hand rapidly along, pressing in the paper, and rolling forward, keeping the right, or basil end of the wire tight, and slightly pulled towards you: the paper round the left end of the wire will have thus become rolled half round the circumference of the wire before the basil end has begun. There is a very great difficulty, at first, in rolling so long a pipe as 17"; it is advisable to begin with the halfpenny pipes which are much shorter; the same wire-former will serve for both. If you cannot succeed, get a person to help you; sixteen fingers can manage the matter easily; but it is best to master the difficulty yourself. Have a tin funnel as nearly the size and shape of fig. 66 as possible. It is best with a lid soldered on the top, with a circular hole in the middle, about the size of a shilling, to prevent the composition from getting spilt. A

27

66

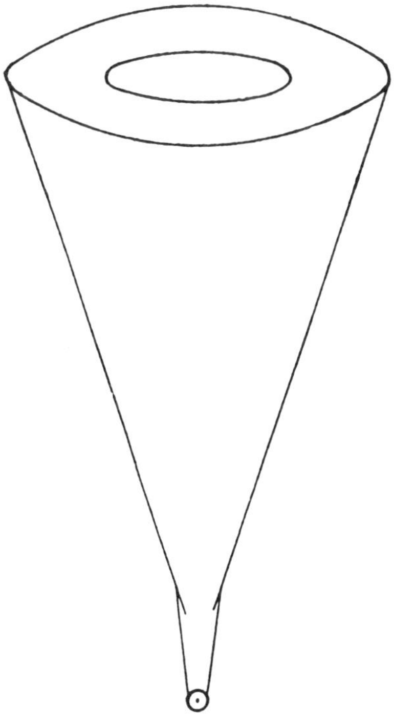

slightly tapering neck is also to be soldered on at the bottom. Take especial care that the mouth of the funnel dips into the neck, and not have the neck sticking up in the funnel. The joint inside must be perfectly smooth. A charging wire will now be required; this should be a square steel wire two feet long, and as large as the neck of the funnel will admit, to move easily up and down; let it also have a piece of lead at the top, to give it weight. Take a piece of brass tube, about 3/8" diameter, and 1" long. Hold one end of the wire in it, with a piece of flannel, or woolen cloth, and pour in melted lead. To charge the cases, first bend a ¼" of the small end of the pipe, over the edge of a knife, into a hook, to prevent the composition running out; insert the nozzle of the funnel in the enlarged end; hold the part of the pipe now round it with the thumb and forefinger of the left hand, slip the wire through the funnel down to the bottom of the pipe, which must rest, for a moment, on the table; then pour in the composition to half fill the funnel, jolt the wire up and down till about 1" of the pipe is filled; then lift the pipe from the table, keep jolting the wire, at the same time turning it round, and let the pipe swing slightly to and fro, till filled. A guide-mark must be filed round the wire, as for squibs, at such a distance as that, when it is level with the top of the funnel the bottom may protrude 1/8". The funnel is to be transferred from one pipe to another without withdrawing the wire. When a number of pipes are charged, see that the large or basilled end, which is the one to be lit, has a vacancy of about ¼"; if necessary, cut it to such length. Take a penholder, taper the end to the thickness of the wire the pipe was rolled on; dip it into meal paste, insert it in the pipe, and turn it round to prime it. Have a length of yellow flax, well waxed with white wax, and wind it tight, 5 or 6 times round the end just primed; cut off the threads to a ¼" in length, and twist them together; or have a piece of wire, as fine as the flax, and wind this round the end of the pipe three times and cut it close: leave the flax or wire, on the pipe. Soak a towel in water; wring it as dry as possible; spread it out flat; lay the pipes in a row, side by side, upon it, like rushes in a chair bottom; roll them up in it, and leave them for 10 or 15 minutes to get damp and pliable. Then wind them upon the usual circular blocks and fasten the end with sealing wax, as at **a**, fig. 261. Be careful that the wax is not in a flame at the moment of touching the case; if necessary, blow it out. Accidents will arise, both from the wax and from the candle, if care is not used. Measure with a tape, round across the wheel, now wound; suppose 4"; cut a piece of blue tissue paper, 4" wide, and any length: cut this

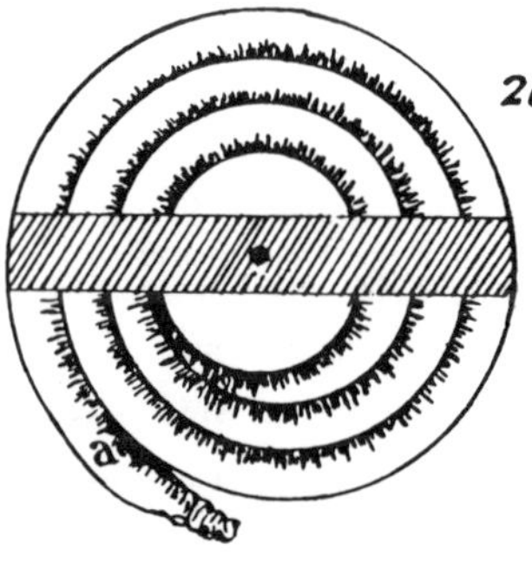

261

across into strips about 3/16" wide, and 4" long; paste a slate all over; lay these strips, 7 or 8 of them, side by side, and press them flat upon it; paste their upper surface; lay one across each pinwheel, and bend it over to make the ends meet in the centre on the other side. If you wish the pinwheel to be of double or triple size, after you have charged one pipe, blow out the dust, paste the small end of a second pipe outside, and stick it in the other. Let it dry; then charge the additional pipe. Observe, the pipes must not lie too long in the towel; if the nitrate gets dissolved it soaks into the paper, which becomes, virtually, touch-paper, and one pipe ignites the other. The wire formers must not be allowed to get rusty; therefore, before laying them aside, smear them with tallow or olive oil, or blacklead them. The wire is much facilitated, too, in delivering, if, just before using it, it is slightly oiled, and then wiped, apparently, perfectly dry. After the wheels are finished, they should be spread out to dry. The flax, or wire, twisted round the mouth, may then be removed, and the pin, or catherine wheel will be complete. The heads of patent pins are now so small, that they will not retain the wheel upon them: French nails are preferable, as they have a large head; see fig. 207; shawl pins are also available; or a straight piece of thin wire, 3 or 4" long, may be furnished with a head made of sealing-wax: make one end of the wire hot, press it against the wax till a sufficient quantity adheres; warm it, and turn it quickly round, and it will form into a round knob, or spherical head.

207

As pinwheel blocks can be obtained in only very few places, if the learner has not a lathe, he may prepare himself a few from vial corks. Cut the cork across with a sharp knife, into pieces rather more than 1/8" thick; make a hole through the centre with a fine bradawl, and burn it out clean with a red-hot wire. To render corks smaller in diameter, cut them round sidewise, the same as you would unwind a case. If provided with a lathe, turn the blocks out of nut hazel. They may also be cut out of the lid of a cigar box, with a button-maker's bit, which differs from a common centre bit, in that both the outer parts are alike and cut round in a circle, without making chips. One can be procured at the tool shops, made to order.

CRACKERS.

A good tough paper for these is 24 mill, 19-lb. double-small hand. A sheet is 30 x 20". Cut off a strip 30 x 4", leaving a piece 30 x 16".

MAKING CRACKERS

71

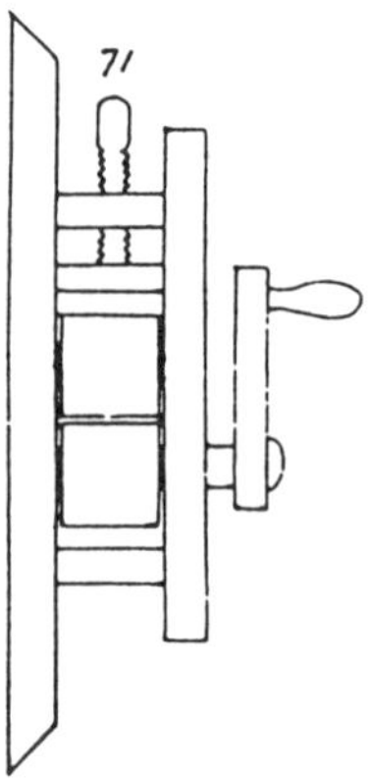

67

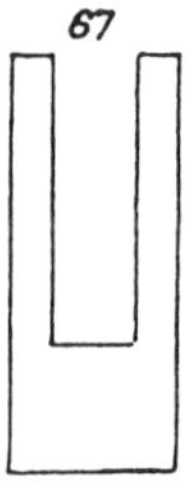

The strip 30 by 4" may be used for squibs, or any odd purpose. For penny crackers cut the 30 by 16" into 8 strips, each 3-3/4" by 16". For halfpenny crackers divide the sheet into 2 pieces, each 10 by 30". Cut each of these into 10, each 10" by 3". For a former have a straight piece of iron, or steel wire, 3/20" diameter, and 24" long. This will not require a basil. Have, also, another wire of half the thickness, for an opening wire. Lay the strips on blotting-paper, as before; and place the wire parallel with the edge of the strip, as there is now no basil. When a number of pipes are rolled and dry, hold one flat on a table; and, with the handle of a knife or tooth-brush rub the pipe flat along, all except the first half-inch, held in the left hand to form a little cup. Now push the opening wire through the pipe to partially open it. Bend the bottom over a knife, into a little hook, as before directed for pinwheels. Take up a quantity; wind a strip of paper round them, as in banging squibs, and pour in a quantity of FFF, or canister powder. Unless the powder is very fine it will not run down. Pipes may be filled with F grain powder, one at a time, by putting a funnel into the cup part, and tapping the pipe with a wire. You can hear the powder run down, and easily judge when the pipe is full. The powder is now to be crushed, till it is half meal and half grain, by rolling the pipes through a plait mill, fig. 71. This is furnished with a wooden screw, at top; the screw presses upon a cross-bit which bears upon two movable collars: these rest on the axis of the top cylinder, by which means the pressure can be adjusted as desired. The cylinders of plait mills are of beech or boxwood; the latter are, of course, the best; but for making great numbers of crackers, it is necessary to have the frames of iron, and the cylinders of steel. If a mill cannot be procured, the pipes may be rolled with a brass tube, or rolling-pin, or passed through an American mangle; but a very good, and far less expensive plan, is to lay them on an anvil, or flat-iron turned upside down, and beat them with a rounded smooth-faced hammer, such as used by shoe-makers; this is a tedious process, but it makes good crackers. The cup end, having served its purpose, is to be flattened, and bent into a hook. Now lay the pipes in a damp towel, like the pinwheels. Take two pieces of pine, each about 6" long and 1½" wide: let the one for the halfpenny crackers be 3/4" thick; the other, for the penny, 1" thick. Saw and chisel a piece out of each, about 4" long and ½" wide, as drawn, fig. 67, so that it looks like the first and third fingers of the hand held straight up, with the second finger, between them, shut. Next procure 20 pieces of wire, each 2" long and about 1/10" thick. The round steel

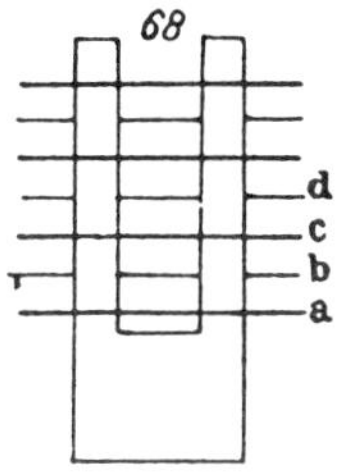

68

wires of umbrellas answer well. Support the cracker-frame in a vise: lay the cracker across the opening; place a wire upon it, as at **a**, fig. 68; bend the pipe over it; lay on it another wire, on the other side, as **b**; bend back, and lay on it the wire **c**; and so on, alternately, till the cracker is bent up. Lift it out of the frame, and let the wires drop. The distance between the wires, in the halfpenny crackers, will be 3/4"; in the penny, 1"; if made according to the directions given. The outside of the frame should be very slightly tapered, or it will be difficult to lift the cracker up out of it. The penny size will probably take 14 wires; the halfpenny 10 wires. Cut the end flush, to expose the crushed powder; wind round it a piece of touch-paper, about 1-3/4" long, $1\frac{1}{4}$" wide; it need not be pasted; fold up the cracker; pass a piece of flax or thread twice round it; twist the thread in and out, backwards and forwards among the folds; and the cracker is complete, see fig. 260. If the paper does not appear sufficiently thick to make a good report, cut the strips wider, and fewer to the sheet; but always cut them up the short way of the paper, or they will not bend properly. The paper for pinwheels and crackers is 30" by 20"; the longest pipe that can be cut is 20": as an experiment, try a pinwheel and cracker the other way of the paper. If the cracker is intended to be put into a jack-in-the-box, shell, or rocket head, push a bradawl up the mouth; have about $1\frac{1}{4}$" of quickmatch; put one end into the vacancy made by the bradawl, bend the other end down to **c**, fig. 260, and wind a bit of flax twice across and round from **a** to **b**, to hold it in position.

260

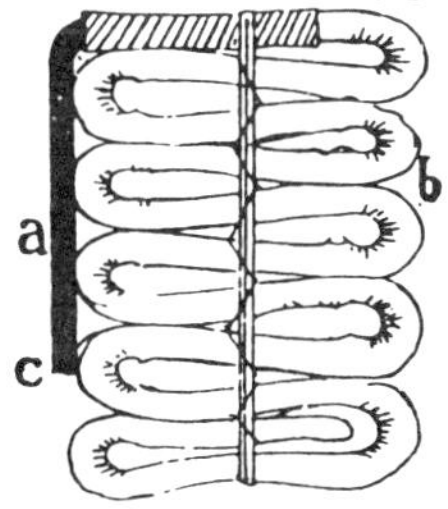

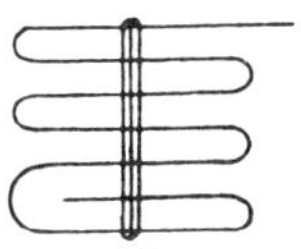

69

The way of bending the cracker is shown at fig. 69, and more clearly at fig. 181; lay the pipe, **mt**, across the aperture; lay the wire **ab**, on it, on the off or farther side; and the wire **cd**, upon it, on the near or hither side; bend the end **m** over to **r**; it will then be of the shape of fig. 182; bend the long end over the wire **cd**, at **s**, this will reduce it to the shape of fig. 183; lay a third wire upon the pipe, on the farther side, above the wire **ab**, and bend the pipe back, which will make it of the shape of fig. 184; and so proceed, alternately forward and backward. It is not necessary that a cracker should contain very much powder. The loudness of the bang depends more upon the thickness of the paper than upon the quantity of the powder. So with maroons; more string and less powder, are better than more powder and less string. A very good paper for crackers is 30-lb. royal cartridge; cut the strips 3" wide; 16-lb. tissue paper may also be used, the strips 5" wide. Crackers for jack-in-the-boxes may be made with brown paper.

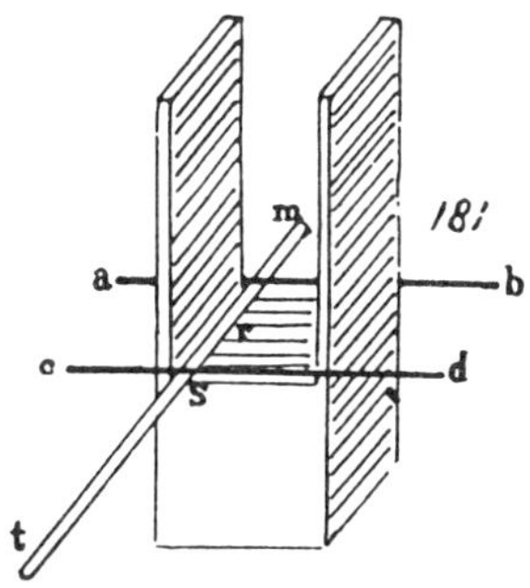

181

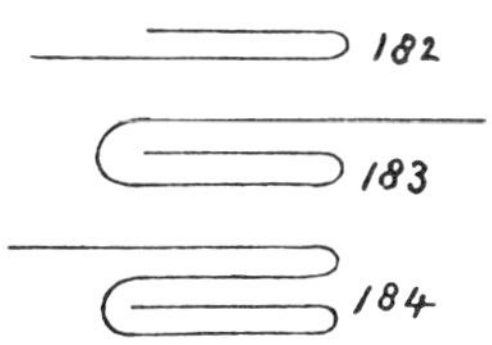

182
183
184

250

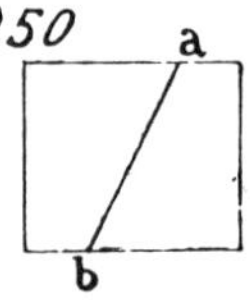

251

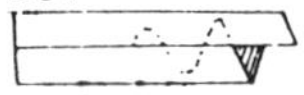

252

To make a cracker with 30 or 40 bangs, it is necessary to join 3 or 4 pipes together, before putting in the grain powder: the mode of making a joining will be understood by attending to the following directions: Cut a piece of paper 3" square, and lay it straight before you. At 1" from the right top corner, along the top edge, **a**, fig. 250, make a mark; at 1" from the left bottom corner, along the bottom edge **b**, make a mark; draw an oblique straight line from one to the other, and cut along it with a pair of scissors. Without disturbing the relative position of the two pieces, draw the right piece a few inches towards the right; paste the farther edges, as usual; lay the wire close along the near edge of the right piece, and roll it up; it will have an external spiral, fig. 252; roll up the left piece in the same manner; it will have an internal spiral, fig. 251; when both are dry paste the external spiral, and screw it, as it were, into the other piece. If managed carefully, and brought up till the edge of the paper forms a straight line, the joining will be as neat as if the paper had not been cut, presenting only a spiral edge, going once round the pipe.

LEADER PIPES.

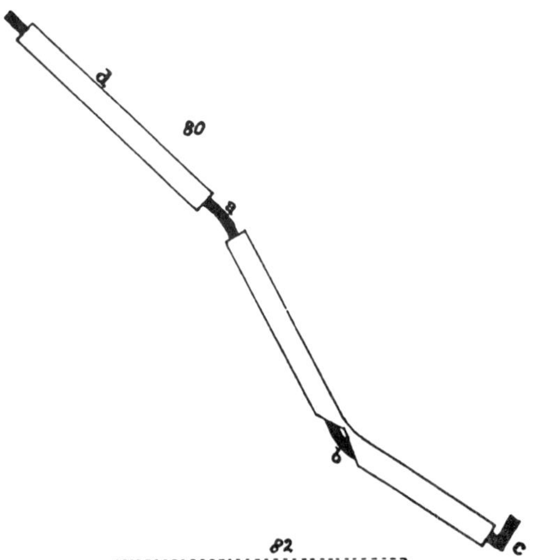

80

82

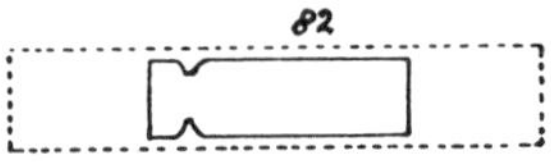

83

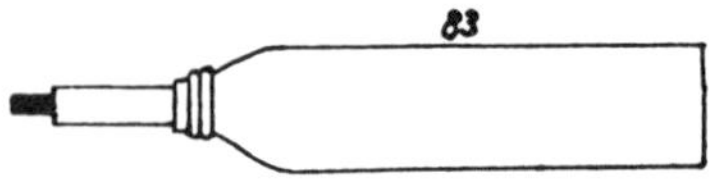

These are for piping quickmatch; they are rolled exactly like pinwheel pipes, on wires of different thickness, to suit the size of the match. They must be large enough to admit the match easily, without much pushing, which would break it; otherwise, the more nearly they fit the match, the more rapidly it blows through. The fire is conveyed from the tail of one case to the mouth of another, by a short piece of pipe, **d**, fig. 80, with the match projecting at both ends; when the fire has to be conveyed to two other cases at once, a fresh piece of piping is slipped on, and the match left exposed, as at **a**; or a piece is cut out of the side, as at **b**; each end should be bent into a hook as at **c**, to prevent it slipping back. Wheel cases are to be enveloped by rolling a piece of tissue paper twice or thrice round them, 2" longer than the case, as fig. 82, shown by the dotted lines. Suppose you have six wheel cases ready charged; lay six pieces of tissue paper flat before you, and work them back with the thumbnail, as before described; paste the 6 edges; lay the 6 wheel cases in a row side by side, and draw the paste brush across the middle, as if pasting your knuckles; then roll one in each envelope, so that it projects an inch at each end. The pasting of the cases across is to prevent the envelope from slipping off when dry. Put the leader pipe in, and tie it with a piece of carpet thread, thin twine, or waxed yellow flax or hemp, as at fig. 83. To

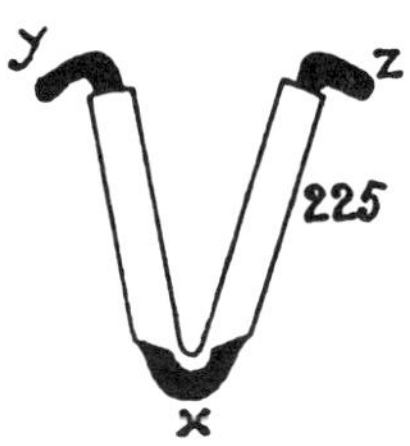

225

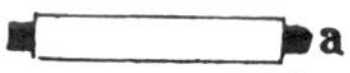

229

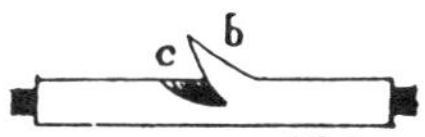

230

join one pipe to another to lengthen it, suppose you have a piece of match 40" long, and two pipes of 20" each, slip both the pipes on the match, so that they touch in the middle; take the end of one, so touching, gather it round the match; pinch the end of the pipe tapering, push it into the other, and bind a piece of pasted paper round to secure the joint. In cutting a matched pipe straight across, of course the match inside gets cut flush; do not leave it so, but, with a pair of pointed scissors, cut away $\frac{1}{4}$" of the pipe all round, and bend the exposed match, as at **c**, fig. 80. If it be required to convey the fire from one case to two others, as in the pigeon, put the exposed match at **x**, fig. 225, into the tail of the first case, and the ends **y** and **z**, into the mouths of the other two cases. It sometimes happens that after a piece is completely leadered, it is desired to connect another piece to it, without disturbing the arrangement. Let fig. 230 be part of a leader of the finished piece, to which it is required to join the leader, fig. 229. Cut a notch in the pipe at **c**; lift it up, like a bird's bill, **b**; insert **a**, fig. 229, press the beak, or bill **b**, down to its original position, and tie round with a thread; secure with a gum-strip. Besides a strong pair of pointed scissors, a small pair, 3" long, with rounded ends, to carry constantly in the waistcoat pocket, will be found convenient for many purposes.

MAROONS.

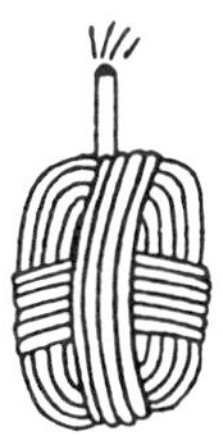

74

Take 3" of a 6/8" or 8/8" rocket case. Fix in one end a cork, $\frac{1}{2}$" long; put in 2" of F grain powder; on this, another cork. Wind string tightly round it, lengthwise, 6 or 8 folds, side by side; bend it to a right angle, and wind 6 or 8 more folds; and so on till covered; then wind crosswise; and again, at right angles, as in fig. 74. Dip it into a mixture of melted glue and red lead, and put by to dry. When dry, with a brass bradawl, filed to a conical point, make a hole through one part, to reach the powder. Make sure of this, by inverting it, and letting a little drop out. Insert a short port-fire, having a piece of match at the bottom, and touch-papered at the top. The best kind of string is laid cord, of which a dozen coils occupy about an inch; they can be counted on the ball when purchased.

Another Method.

Take a 5/8" rocket case, 5" long and unchoked. Put in one solid inch of plaster; upon this pour in 3" of grain powder, on this put another solid inch of plaster. When dry, wind string tightly round it, up and down the cylindrical part, till of two or three thicknesses. Paint the string over with melted glue and red lead. Make a

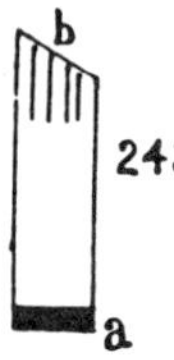

243

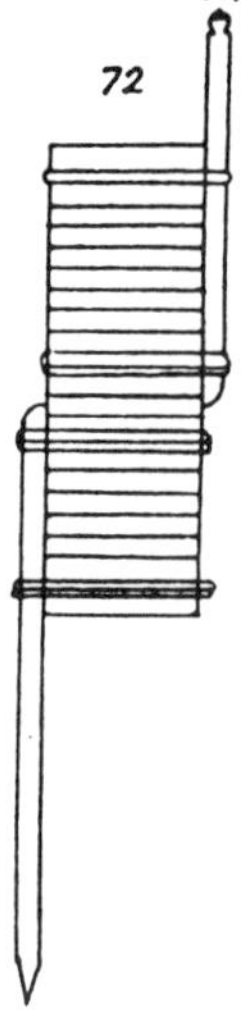

72

bradawl hole in the middle of the side; fasten to it a squib, rammed with two inches of powder, cut sloping at the lower end like **b**, fig. 243, and having a long piece of quickmatch carried down the case, to enter the hole. Cover the joining with two layers of pasted paper. On the other side below, fasten a short piece of a rocket stick, with a point, to put into the ground, fig. 72. After lighting the touch-paper, remove to a distance; as bits of string are likely to get driven into the face, on the explosion. A maroon of this shape may be enclosed in a rocket head; simply insert a bit of match in the hole, and bend it down the side. Proportion the size to the rocket, but have an inch of plaster at each end. Even a squib case treated this way makes a good report.

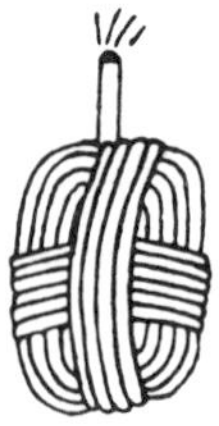

74

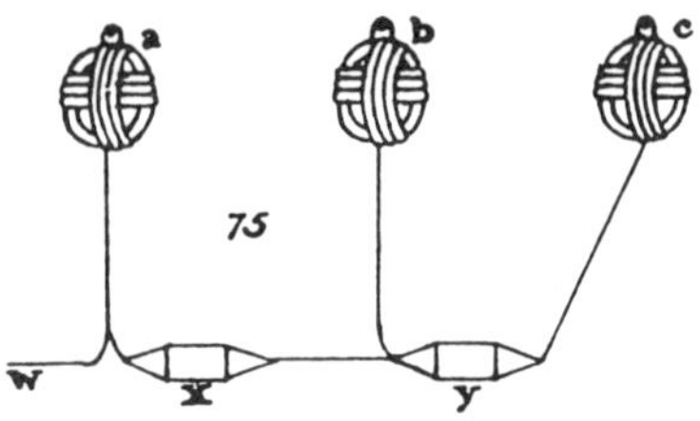

75

To fire a salute with maroons, at regular intervals of time: Charge a port-fire, and saw it into inch lengths; envelop each in a piece of tissue paper, 3" wide, and long enough to go twice round the port-fire. Hang the maroons to hooks, or otherwise suspend them on a frame, a foot distance from each other, as **a, b, c,** fig. 75. Underneath them fasten with binding screws, or tie to nails, the port-fires **x, y.** Connect the port-fires with one another, and with the maroons, by leaders, in the usual way. On lighting at **w**, the first maroon explodes, and the first port-fire catches; this, having burnt, lights the second maroon and the second port-fire; and the port-fires being of the same length, the intervals of time between the explosions of the maroons will be the same. The match is to be carried straight into the maroon, as shown in fig. 75; the port-fire in fig. 74 is simply to allow time to remove to a distance, before the maroon explodes.

GOLD AND SILVER RAINS.

These are little unchoked cases, 2½" long, rolled dry, on the squib-former, to put into rocket heads, mouth downwards. Cut a piece of writing paper 2½" square, and a piece of tissue paper 2-3/4" square; paste or gum the edge of the tissue paper, and roll it outside the writing paper, so that it shall be flush at one end and overhang the other ¼"; tuck in this overhanging part, to form a bottom; charge with the funnel and wire; and prime the mouth with very slightly damped meal. Or, instead of this way, cut a piece of 60 lb. kraft paper 4" long and 1-3/8 wide; paste it all over, and roll it on a 3/8 former, so as to make the case 1-3/8" long. When dry, cut a notch in the lower end, as in fig. 61. Take an inch of quickmatch, and bend it into the shape of the letter V; set it upright, and slip the

61

case over it, so that it shall stand half in the case, and half out, through the notch. Put in a little chlorate meal, slightly damped with lac solution; on that a small scoopful of gold rain; insert drift, and mallet both together; fill up with gold rain, and close with a film of plaster. The case will thus be primed with ½" of protruding match, held in firmly by the dry chlorate meal. These rains may be put into the rocket head in two tiers, one above the other.

PEACOCK'S PLUMES.

These are a combination of rain and star. Roll them like pill-box cases, on a 3/8" former, about 1-3/4" long; charge one end 3/8" deep with coloured fire, driven in dry; fill up with gold or silver rain, with a film of shell fuse at the top to bind. Cut a bit of match 2¼" long, lay it outside the case, so that it projects 3/8" at one end; envelop it in a piece of tissue paper, fig. 90; tuck in the paper, to press the match, at one end, on the colour; twist the other to a point. Both ends thus light at once; and the rain appears, like a coloured star with a tail. A mixture of enveloped coloured stars, and naked streamer, or steel stars, make a good imitation. Blue No. 10 contrasts well.

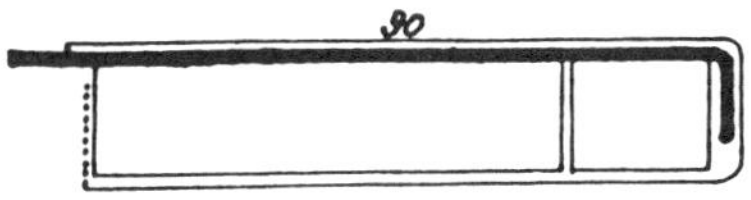

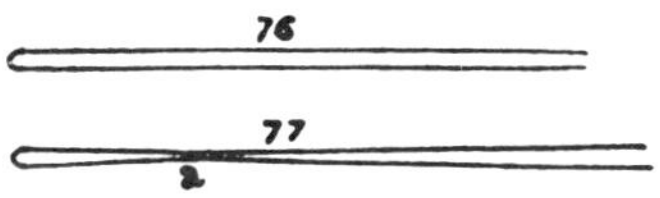

To prime a case with match laid flat on the mouth. Take a piece of thread, or fine string, and fold or bend it in the middle, as at fig. 76. Tie a knot near the bent end as at **a**, fig. 77. Bring the knot **a**, up to the side of the case, as at fig. 78; pass the loose ends round, and tie in a knot, at **b**. Lay the bits of match flat across, as at **e**, fig. 79; bring the threads together, and tie them at **c**. Or bend tape match across, and tie as at **d**.

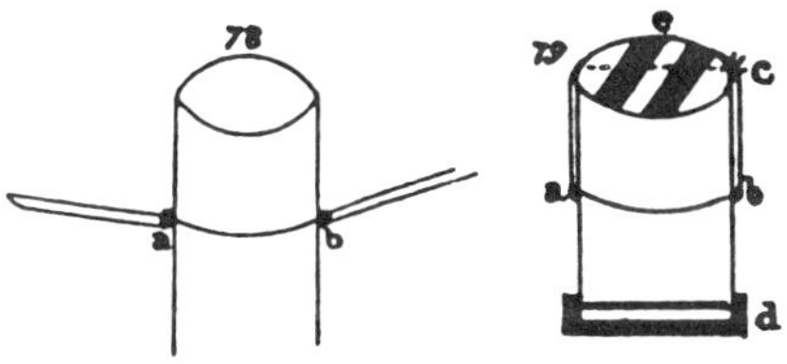

To get a fine thread through a long pipe, or the hem of a bag. Take a piece of copper wire, and bend it round at one end, as **z**, fig. 88. Pass the end, **z**, forward, and push it through to the other end of the bag, etc., then bend it to the form of **x**, fig. 89; pass the string through the loop **x**, and pull the wire back.

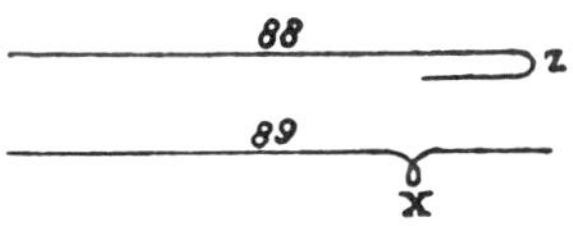

SAUCISSONS.

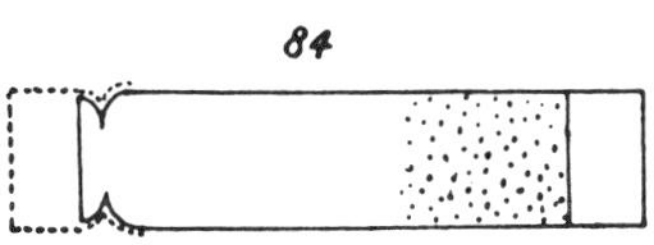

These are a large kind of serpent, charged on a nipple, like a wheel case, with solid drift and mallet. They may be 5/8" or 6/8" or larger: about 3½" long. Drive in brilliant fire, or gerbe, fig. 84, 1½"; fill up to within ½" of the top with F grain powder; and plug the end with plaster of Paris, or a bit of wood fastened with a tack or two. Press a piece of touch-paper, or tissue paper, into the shape of a deep

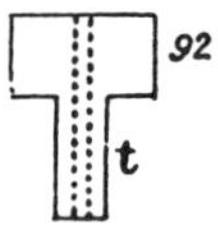

pill-box; fill it with F grain powder; fit it to the mouth of the saucisson; tie round the choke; brush, with meal paste the bottom, and dip into dry meal. These saucissons are to be fired in a volley. Procure, say 2 dozen, iron tubes, **a, b, c**, etc., each a foot long; fit them with a wooden bottom, fig. 92, having a tenon, **t**, an inch long; let it be fastened with a screw on each side. Bore a hole through, to make a communication with the mortar formed by the tube. Take a board 1" thick, of suitable length and breadth; bore in it 2 dozen holes, of a size to fit the tenons; glue these in, so that the tubes, or mortars, stand upright, in rows, side by side, like the pieces on a chess-board. Invert it. Nail a rim all round, so as to make a box 2 or 3" deep. Cut a groove from hole to hole of the tenons; connect all the holes with naked match, also push a bit of match up all the holes in the tenons; now fill the box with sawdust, pressed in as tight as possible, and nail a board on to serve for a bottom and to keep the sawdust in. Invert it; and put a saucisson, mouth downwards, into each mortar. Fig. 84 represents a single saucisson; **w, w, w**, fig. 85, saucissons in the mortars. On firing, it is evident the cases will be driven out rapidly, one after the other. The sawdust prevents the flash igniting the whole at once.

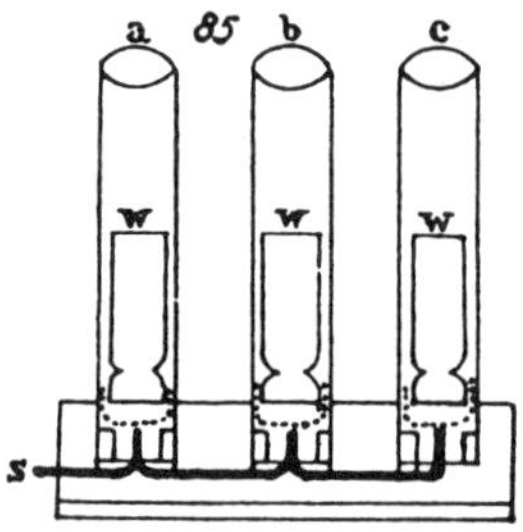

Saucissons, instead of being charged with grain-powder, to cause a report, may be fitted with a cylindrical head, containing small stars, short gold or silver rains, serpents, etc.

PEARL STREAMERS.

Paste brown paper all over, and roll up a case of four or five thicknesses, on $1\frac{1}{4}$", or $1\frac{1}{2}$" former, like a rocket or other case; when dry, cut it in the lathe (see fig. 27) into inch lengths--inch-deep bottomless pill-boxes. Set one on a foot, fig. 9, to enter about 1/8"; mallet in the pearl streamer powder, till nearly full, then a little meal powder; remove it from the foot, and into the void left press in flat, with a knife, a little plaster of Paris, to form a bottom. They will have the appearance of bungs; fire them in volleys, like saucissons, from suitable-sized mortars; primed end downwards, of course. Match may be tied on, as in figs. 78 and 79.

BLUE LIGHTS & STAR CANDLES, OR STAR LIGHTS.

These are little cases, charged with the funnel and wire; the latter are filled with spur fire. They may be made as directed for gold and silver rains, only five or six inches long, with writing paper and tissue paper.

PRINCE OF WALES' FEATHERS.

These are simply pin-wheel pipes, usually of coloured tissue paper, charged with pin-wheel powder, and not wound on a block, but kept straight.

LANCES.

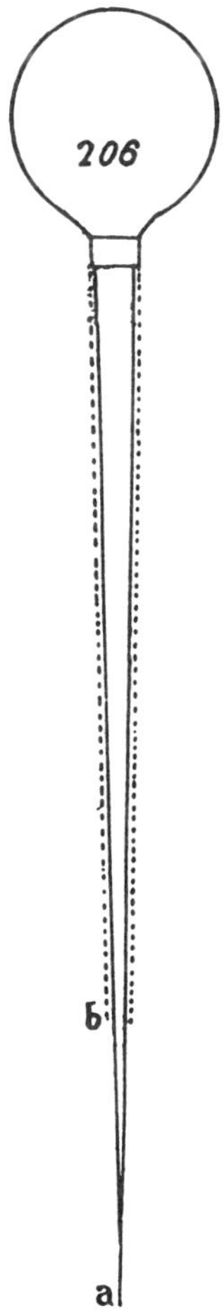

248

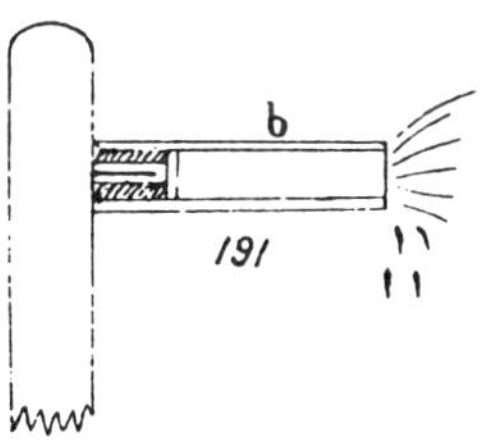

These are little cases charged with white or coloured star composition. They should be of white or coloured tissue paper paper, rolled dry on the squib-former, and secured at the edge with paste in the usual way. They may be from $2\frac{1}{2}$ to 4 or more inches long, as may be required. Cut a sheet of thin tissue paper into 80 equal parts (2, 2, 4, 5), each piece will be 3-3/4" by 2". Cut writing paper into pieces $3\frac{1}{2}$" by 2". Roll the writing paper round the squib-former, beginning with the edge of the paper, so as to make the case $3\frac{1}{2}$" long, and of uniform thickness; and the tissue paper, with its gummed or pasted edge outside, flush at one end, and projecting $\frac{1}{4}$" at the other, to tuck in and form a bottom. Charge with the wire and squib funnel, fig. 65, till full within 1/8". Prime the mouth by pressing it into slightly damped meal. Lances containing strontium nitrate; or, indeed, any lances, may be primed with slightly lac-damped chlorate meal. If the cases are made longer, they must not be made thicker, as they are intended to burn with the compositon, without deteriorating the colour; two folds of writing paper, and the same of tissue paper, are sufficient to keep them straight, which is all that is required. The following colours may be tried for lances:--White, 1, 10, 12; sugar blue, 1, 4; blue, 2, 3, 17; red, 3, 8, 10, 23; green, 3, 6, 11, 21; yellow, 1, 3; mauve, 1, 5; purple, 3, 10.

Lances are used for forming letters or designs, similar to gas jets. Frames for letters and devices are made with thin pieces of pine for straight lines, and cane or hooping for curves. A number of wires, 3" apart, are driven into the frames, and the lances are fixed on the wires, so as to project horizontally, like **b**, fig. 191, in order that, when burning, the dross, or scoria, may fall, and leave the fire clear and bright. The following is the way or proceeding:--Procure some inch French nails, fig. 207. Drive them into the frame *about halfway home*, so that $\frac{1}{2}$" is left projecting; then, with a pair of cutting pliers, or nippers, cut off their heads. Now make a tapering bradawl. For this purpose, procure an upholsterer's needle, fig. 248; break off $2\frac{1}{2}$" of the point end, and fix it in a handle, so that 1" shall be left protruding; or, instead, take a stiletto, or printer's bodkin, fig. 206, and shield it, by pasting a strip of paper round it, so as to leave 1" at

249

209

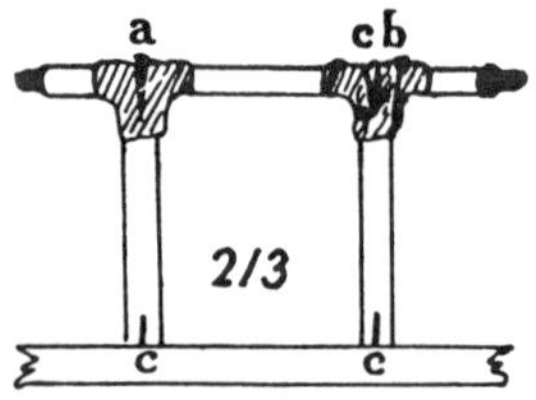

the point end protruding. Next procure a sail-needle; fig. 249, about $2\frac{1}{2}$" long; it may be obtained at any tool shop; the point is triangular and excessively sharp. Fix it in a handle, so that only about $\frac{1}{2}$" is left protruding; a pine handle is best; if a beech bradawl handle be used, a hole must be carefully bored up it. Fix the needle, eye end uppermost, upright in a vise, put the handle over it, and drive it down with a few gentle taps. To fix the lances, push the pointed bradawl, or stiletto, 1" up the bottom of the lance; set the lance upon one of the wires, or beheaded nails, **c**, **c**, fig. 213, and pull the bottom part of the lance forcibly down over the wire; this will cause it to bite, and hold firm on the nail; it will hold firmer still if the nail be rusty, or if the bottom of the case be previously touched with a dip of glue, or gum-water of the consistency of pudding. Having completed the letter, or device, proceed to leader it. For this purpose have a supply of 3/8" rivets, fig. 209, but shorter: they are about 5*d.* per lb. Take a length of piped or leadered quickmatch; lay one end of it on the mouth of a lance; push the triangular-pointed sail-needle through the middle of the match down into the lance, and turn it round to insure the breaking of the priming; withdraw the needle, and push in a rivet, **a**, fig. 213; and so proceed till the device is leadered. Now take a gum-strip, or a piece of pasted tissue paper, $\frac{1}{2}$" wide, and $1\frac{1}{2}$" long; press it over the joining, and smooth it round the lance, so that it assumes the shape of the capital letter T, as drawn. The match is thus bradded to the lances, and prevented from slipping off, by the gummed or pasted strip. Another way of matching designs, is to fasten the leaders to the lances with small pins, patent short whites. Push the pin through the *side* of the leader down into the *side* of the lance, as at **b**, fig. 213; then push the sail-needle through the *middle* of the leader down into the *middle* of the lance, as at **c**, to make a connection, turning it round to break the priming, and secure the leader with a gum-strip. A common round bradawl will not answer the purpose so well as a triangular one, as the sharp edges of the latter break the match and priming, and insure the ignition.

TO FORM A DEVICE OR DESIGN.

Take a sheet of paper, and draw upon it a representation of whatever is intended: as a temple, a mosque, a tree, a ship, a horse, a peacock, a portrait, George and the dragon, the lion and unicorn, fig. 247, etc.; then cross the design with lines, at regular distances, and at right angles, so as to cover it with squares, as fig. 86. It is now requisite to have a floor, of a considerable size; but not

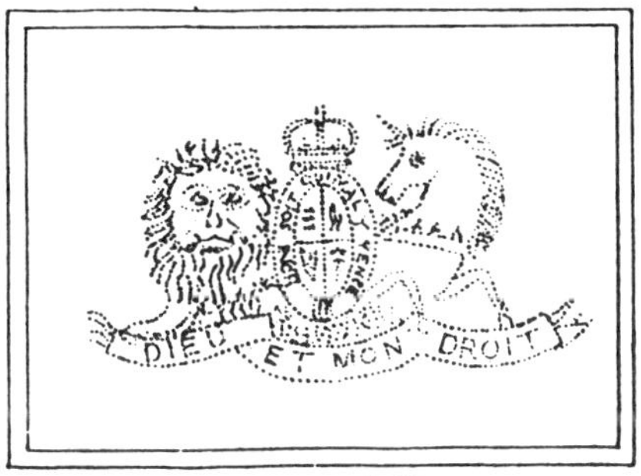

247

86

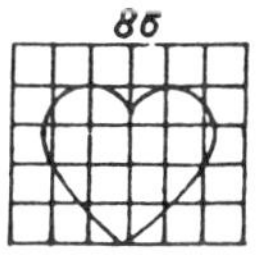

87

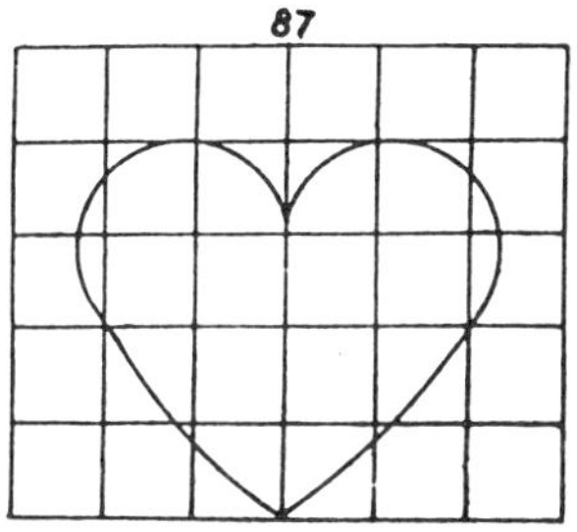

91

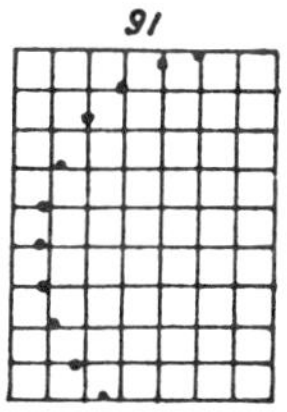

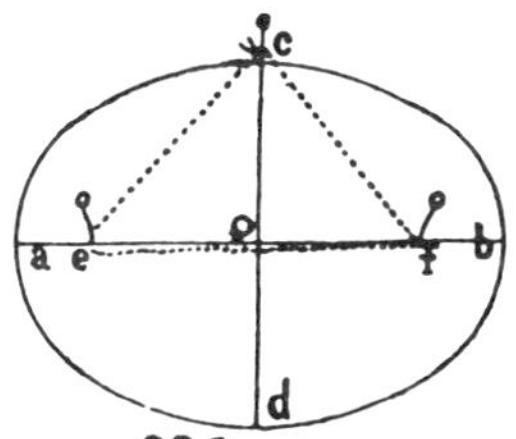

226

necessarily so large as the design intended to be fired, as a part can be done at a time; this floor must be divided into large squares, and the device from the small pattern, fig. 86, transferred to it, and the outlines chalked on the floor, fig. 87. After this a number of frames are to be made, of pine, or other wood, like square lattice-work, as fig. 91. A frame is then to be laid on the floor, so as to cover *a portion* of the design; and the French nails, before mentioned, driven in, at distances of 3", to receive the lances. Every frame must be numbered, and a copy kept in miniature, so that they may be correctly fitted together, to form the figure without delay or error. Number the first row A1, A2, A3, etc.; the second row B1, B2, etc.; running the numerals horizontally along the top; and the letters vertically down the side of the design. To distribute the fire rapidly over an extensive piece, it is necessary, at points, to make one leader light several simultaneously. Suppose 6 are to be lighted at once: bring the 6 ends together and tie them; envelop them with a piece of tissue paper, projecting a couple of inches; in the void put a scoop of meal; bring in the single match that is to light them, and tie as usual. Paste a piece of paper over all to make secure.

In devices, an ellipse is sometimes introduced; the proper way of forming it is the following:--Lay a sheet of writing paper on a smooth pine board; draw upon it the length and breadth, the transverse and conjugate diameters, **ab**, **cd**, fig. 226, bisecting and crossing each other at right angles, in the point **o**. With a pair of compasses, from the point **c**, with the radius **ao**, mark the focal points **e** and **f**; drive in 3 pins at **c**, **e**, **f**, slightly sloping outwards, pass a piece of thread round them, as shown by the dotted lines, and tie it in a knot at **c**; then pull out the pin **c**, and work a blacklead pencil round, within the thread, keeping it stretched. To make elliptical flower beds, drive 3 stakes into the ground, and use string.

TO PRESERVE STEEL FILINGS OR CAST IRON BORINGS.

Put 1 lb. into a frying-pan, or iron ladle, with 3 ounces of marine glue; set it over the fire; and, as the glue melts, stir it about till thoroughly incorporated with the filings, or borings. When cold, bottle them, and cork. Marine glue is made by putting pieces of india-rubber into mineral naphtha; the caoutchouc swells up; it is then triturated into a uniform mass, mixed with shellac, and melted. It is easily made. Wax solution, or stearine solution, rubbed up in a mortar with steel filings or cast-iron borings, also preserves them. The nitrate of the gunpowder, however, attacks them, and ultimately rusts them; so that fireworks are never so brilliant as when recently charged. To try whether filings are good, sprinkle a pinch of them into the flame of a piece of lighted brown paper. The effect of brilliant composition is seen in the grinding of knives and scissors; or gerbe-powder in the burning of white-hot iron, as it is taken from the fire by the blacksmith.

SHELLS.

CHARGING SHELLS

Shells are hollow paper globes, fired vertically, from mortars, or iron tubes. They are made of various sizes, from 3" in diameter to 16". To make a 3" shell. Turn a wooden ball, 3" diameter; and, round the middle, that is, the equatorial circumference, cut a V groove, or triangular channel, deep enough to receive a piece of raw or naked match. Remove it from the lathe, and cut it into two halves at right angles to the groove, that is, round a meridional circumference. Construct a pine box, 4" square, 2" deep. Place one of the half globes, flat surface downwards, on the middle of the bottom of the box, and secure it with screws from underneath. Brush it, and the inside of the box, all over with sweet oil, with a camel-hair pencil. Put some water into a basin; sprinkle into it as much plaster of Paris as judged necessary; about 4 tablespoonsful; pour off the water which floats above; stir up the plaster till homogeneous; pour it into the box; and, with a sash brush held upright, beat the plaster in with the points of the bristles. Leave it to set.

Instead of having a wooden ball turned, a hemispherical concavity may be made by pressing, half way, into sand, one of the painted india-rubber balls sold at the toyshops; and pouring plaster over it. Or a basin, an inch diameter larger than the intended shell, can have the plaster mixed up in it till about three-parts full; and then the bottom of an oil flask can be pressed into it. A narrow strip of blue paper should be previously pasted round the oil flask at the proper height, as a guide to know the proper depth to which it may be pressed. The plaster, when partly dry, must be neatly trimmed; and may be left permanently in the basin. Or a stiff paper cylinder, or a tin cylinder, may be made, an inch larger in diameter than the diameter of the intended shell: put the oil flask into this, neck downwards, and pour in dry sand, till only the hemispherical bottom of the flask is left exposed; level the sand, oil the flask; and pour in plaster, as before. Be careful that the mould is not less than ½" thick in any part. Or one or two, or more halves of the zinc or copper globes, used for ball-taps, may be obtained of the plumber, and used for moulds, without further preparation.

To Make the Shells.

Procure two kinds of paper; one kraft, the other sugar paper, printed paper, paper hangings, or any paper of a different colour from brown. The shell is supposed to be 3"; half as much again is 4½"; add to this 1½" for a flange, gives 6". Cut out a circular piece of the brown paper, 6" in diameter. Snip it all round with the scissors, in slits, reaching from the circumference, half way towards the centre: soak it in water, and lay it on a towel to drain. Have a piece of sponge, abcut the size of an orange; soak it also, and wring it. Grease the mould, to make it deliver. Place the piece of brown paper in the plaster concavity, and press it in neatly with the sponge, in all directions; it will fill up the mould, and overhang all round; press the overhanging part flat, so that it forms a flange. Cut a piece of another-coloured paper, into a strip about 2" wide, and paste it well on both sides; then cut it across, backward and forward, like the letter W, which will reduce it to V-shaped triangles. Take these up, one by one; lay them in the brown paper, pasted side downwards, and press them neatly in with the sponge, or with the fingers, making each slightly overlap the other. Let these also overhang to thicken the flange. This being done, proceed with a layer of brown; and so on, alternately, till it gauges internally about 2-3/4" across, which is best ascertained by having a solid globe, or mould, to press in till it fits. The shell will now be about 1/8" thick. Remove it, and proceed with the others. The different-coloured papers enable the eye to detect whether any part is left uncovered. When dry, lay the half shell flat on a board, and with a sharp knife, such as used by shoemakers, cut off the flange. Have a cylindrical chuck, and turn the end hemispherical, so as to be a tight fit in the shell; chalk it, and press the shell firm on it to make it grip, then cut it with the chisel in the usual way. Also cut a hole, 9/16" diameter, in the middle of one hemisphere, to receive the fuse; to it glue another hemisphere; and cover with two more layers of paper, or pieces of calico. The shell may then, when dry, be filled with any kind of stars or rains that can be got through the hole. Along with the stars put 1/16 of their weight of mealpowder for a bursting charge: that is, if the stars weigh 4 ounces, as they probably will, put 4 drams of mealpowder. Charge a Roman candle case, 3/8" internal, 9/16" external diameter, with shell fuse (see fig. 59); saw it into 7/8" lengths. Cut a piece of calico 2" wide, and long enough to

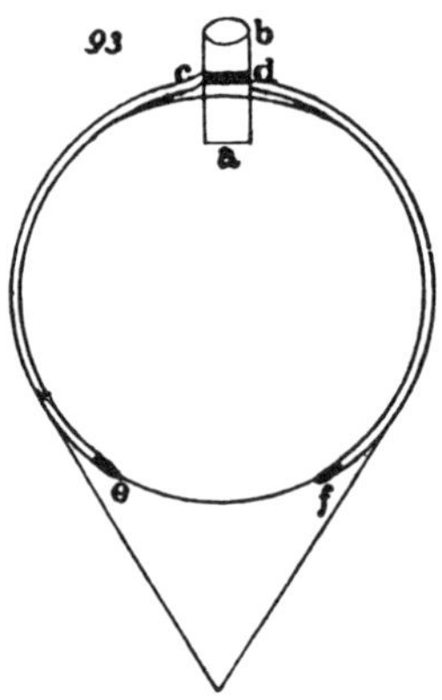

go twice round the fuse; paste the calico all over, and roll the fuse in it, so that the envelope is flush at one end, and overhangs 1" at the other. Glue this in the hole of the shell, the flush end, **a**, fig. 93, of course, inward: the enveloped end, **b**, being to receive the match. 3/4 of the fuse may be pushed in, the rest left to project. With a printer's bodkin, or a stiletto, make a hole through both sides of the envelope, as at **c** and **d**; these holes are to be in a line with the groove; put through them a piece of raw match, in such a manner that it shall lie across the mouth of the fuse, and go in the groove round the shell underneath, and reach to **e** and **f**. Paste strips of paper over the match, the same as with tourbillions and saxons; or the part of the match lying in the groove may be in a leader pipe. Weigh the shell, and take about 1/5 of its weight of coarse grain powder, for a blowing charge. Make a cone of two or three thicknesses of paper; put the blowing-charge into it; stick the cone on the bottom of the shell, and set by to dry. In the calico mouth of the fuse, tie a long piece of leadered match, and paste a strip of paper round to make secure. If the mould has been made with the indentation of an oil flask, or with the globe of a ball-tap, it will, of course, not have a groove to receive the match but this is of little consequence; the match can be drawn round outside, and covered, so as to appear as it does on tourbillions and saxons, like a vein on the back of the hand, when the fingers are held downwards. In filling up large shells the blowing-powder is enclosed in flannel or serge bags; which may be conical, like the paper ones; or cylindrical, and tied at the bottom with a string.

Formerly mortars were made of sheet iron, riveted and bound round with cord, which latter would not prevent them cracking, if they were not thick enough; they are now made on an improved principle: the iron is rolled, by powerful machinery, of three thicknesses, exactly like a squib case; it is then made white-hot, and the three are welded together, with a steam hammer. Large mortars, also, have an iron bottom, or breech fixed in them, and are farther strengthened with a couple of rings, put on hot, and shrunk by cold, like tires on wheels; a third ring is put over the other two, as in the Armstrong guns. The mortar is placed in a hole dug in the ground, a few inches left standing out; the earth is shovelled in and driven down firm; a penthouse lid, to keep out rain, dirt, and insects, renders it complete. A small mortar may be a tube, open at both ends, and fitted with a wooden bottom, to which it is to be firmly screwed, a hole on each side having been previously drilled

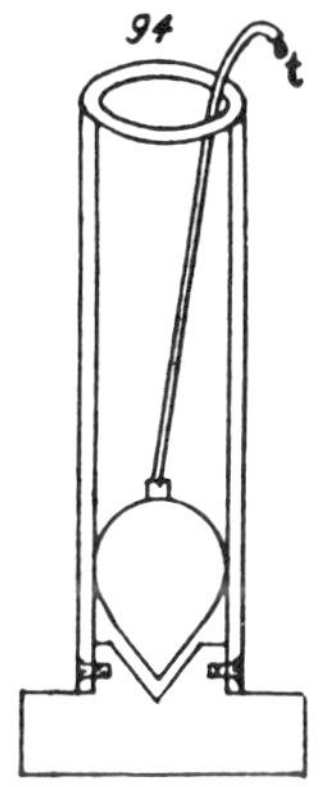

to receive the screws. Fig. 94 represents such mortar: it may be 4 diameters high, and the foot should have a conical hole turned in it to receive the cone fastened to the shell. A paper mortar rolled like a case, till $\frac{1}{4}$" thick, will serve for several times, it should be bound round with strong cord. The match is lit at **t**; but this may have a long bit of touch-paper attached to it if preferred, in order to be able to remove to a distance.

Instead of making a plaster mould to form the shells in, the shells may be made by covering a wooden sphere, with paper, on the outside; when dry, they may be cut round in the lathe (a cross mark, with a pencil, having previously been made, as a guide to bring the same parts together again); the wooden mould removed; the cut edges glued; and the shell fitted up, in the usual way. Clean oil flasks may be covered with six or eight thicknesses of paper; paste an inch, or two, round the neck; when dry, cut through the cover, near the spherical part; file a notch all round, and snap it off. I have made excellent shells this way; the chief objection against them is their limited size. Another ready way of making shells, is to cover the india-rubber air-balls, of the toyshops, almost as thin as soap-bubbles; when the cover is dry, a hole may be cut for the fuse, with a penknife, and they are ready at once to receive the stars. Their shape is that of a prolate spheroid, fig. 106. After eight thicknesses, or more, of paper have been pasted on, measure with a tape round the equatorial circumference, **b e d**, which suppose 17"; add 1 to this = 18". Measure from the pole, **a**, down the meridian by **e** to the opposite pole **c**, suppose 11". Cut a piece of tissue paper, 18" by 11"; fold it down the middle, to a double thickness of 9" by 11"; fold again to 4-1/2 by 11; there will now be eight thicknesses. Pencil the shape fig. 107 upon the top, and cut through the whole. Paste the eight gores on, as in fig. 106; for ornament, half may be pink; half green.

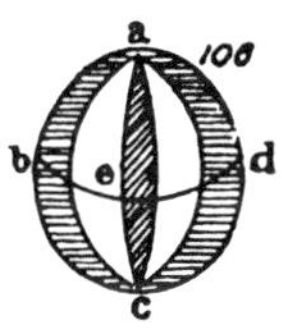

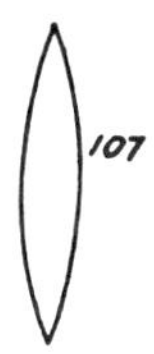

Cylindrical Shells.

A sphere is, by no means, the best shape for a projectile; no one would think of making a Roman candle star like a marble; the great Minie bullets and the bolt-shot for our great guns are cylindrical, and far better fitted than globes for straight and rapid flight. Why not adopt the same shape for shells? I have fired cylindrical shells, and find them the answer. Have a former for a small one, $2\frac{1}{2}$" diameter and about 6" long, and with a handle like fig. 25. Roll a case upon it 5" long, till the outside gauges 2-3/4", or a trifle more.

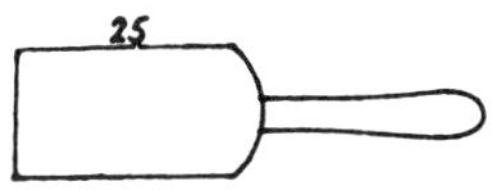

95

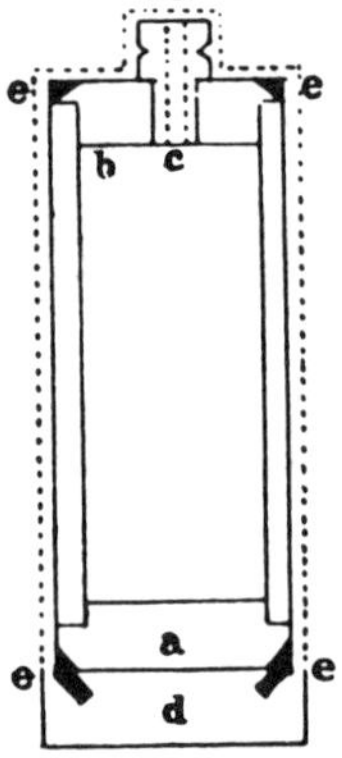

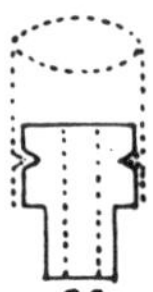

96

105

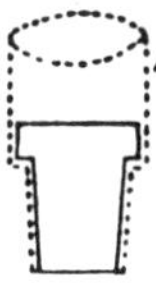

104

Turn a wooden bottom, 1½" thick: half of it a tenon to fit the inside of the case, and half of it a flange, equal to the external diameter, **a**, fig. 95. Glue this firmly in and further secure it with 4 inch French nails. The top may be of the same shape; an inch thickness will be sufficient. A hole, 1" in diameter, may be bored in it with a centre-bit, to receive the fuse. This fuse may be a cotton reel, with one of the flanges sawn off, the central hole enlarged, and the end filed or turned slightly tapering, as fig. 105; the enveloping piece of calico, or glazed lining, can then be passed to the bottom, as indicated by the dotted lines; but a better way is to turn a piece of hardwood to the shape of fig. 96, with a collar, to prevent its being blown through, and a groove, by which the calico envelope can be tied. The length of the fuse may be about 7/8"; charge it by putting in very little at a time, and well mallet it. Pass a piece of naked match over the mouth of the fuse, and down the sides of the cylinder, as indicated by the dotted lines, fig. 95, having previously cut out a notch at each corner, top and bottom, **e**, **e**, **e**, **e**, to guide it in a straight line. Take a piece of tissue paper, about 20" long and 7" wide; paste it all over, and roll the cylinder in it, in such a way that it shall be flush at top and project at the bottom; press it round smooth with a cloth or sponge, till the leaders form a vein on each side. When dry invert it; put in the blowing-powder; press the ends over neatly to form a bottom; and cover the whole with a circular piece of pasted paper. A shell of this size will hold forty gold rains, and a score or more blue stars, which have a very pretty effect. The head and the fuse may be in one solid piece, if preferred, as fig. 104; if the hole is longer than 7/8", only 7/8" of it must be filled with powder; a piece or two of match may fill up the rest. The mortar for this shell should be 18" or 20" high, and 3" internal diameter. The shell, when fitted up, will probably weigh about 14 or 15 ounces. The blowing-powder may be 2 or 2½ ounces of F grain, according to the fit in the mortar. The hole in the wooden fuse should be 3/8" diameter. See that the match is everywhere carefully covered. A shell of this size will be amply sufficient for amateur purposes. Perhaps it will be advisable to try a fuse 6/8" long the first time; it will be better that the shell should burst while still ascending, than that it should pitch. One or two trials are, in all cases, necessary.

A cylinder holds half as much again as a shell of equal diameter; consequently, when the depth is twice the diameter, which is the best proportion, it holds three times as much as the spherical shell.

If the latter be 3" diameter, it may be represented by the numbers 3 x 3 x 3 = 27. If the cylinder be 3" diameter, and 6" deep, it will be as 3 x 3 x 6 x 1-1/2 = 81. Practically, it will hold more, especially of rains or serpents, as they pack better in a cylinder.

The spheroidal shell, like the spherical, is also 2/3 of its circumscribing cylinder.

Shells, for war purposes, are both spherical and cylindrical; the latter are sometimes built up in pieces, which dovetail into each other; they are then set in a cylinder, and melted lead is poured round to bind the pieces together. For lighting up the country to discover the movements of the enemy on a dark night, shells are fired, containing strong calico parachutes, carrying blue lights. Magnesium lights are, of course, more effective.

102

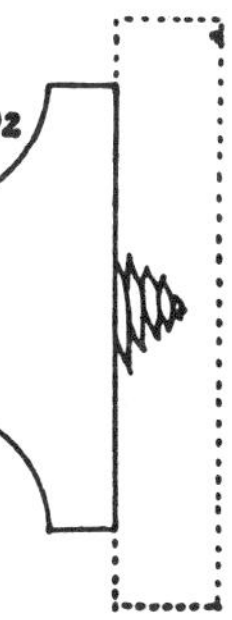

The bottoms and tops of the cylindrical shells, previously described, also the bottoms of mines, may be turned out of a plank of wood--elm, ash, or mahogany; and as many amateurs who possess a lathe, know of no cheap contrivance for holding flat pieces of board without making a hole through them, the following plan may be adopted. On the screw-chuck, fig. 102, screw a piece of pine board, previously sawn nearly circular, and as large as the lathe will take: let it be of a thickness to entirely hide the point of the screw; turn it to a circle, and over the face of it, with a blacklead pencil, while it is revolving, make a number of concentric circles, as in fig. 103. Also cross it with two straight lines, passing through the centre, at right angles to each other. Cut the wood intended for the bottom of the shell or mine, into a square; make a hole at each corner, and with 4 screws or 4 French nails, screw or nail it on the face: the concentric circles, and the two diametrical lines crossing them, will act as guides to centre it. I have thus been able to cut into a circle, in the lathe, so thin a substance as a piece of writing paper. It may be held on with 4 tin-tacks, or a touch of gum.

103

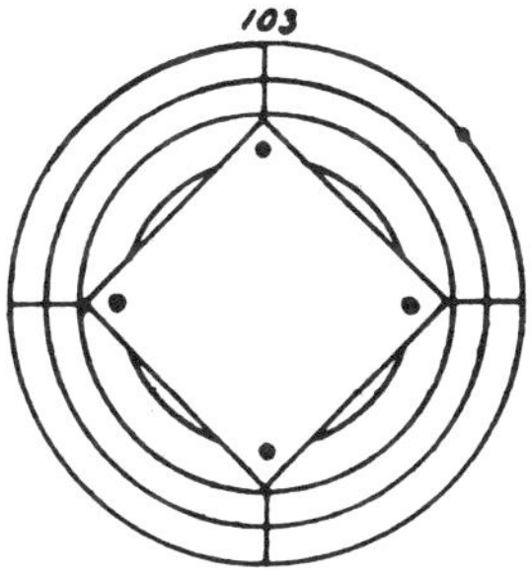

A nest of 6 or 9 little shells, each containing a separate colour, rains, serpents, crackers, etc., may be enclosed in a large shell: the fuse of these may be a very short piece of a squib-case, with the string wound 7 or 8 times round the choke to form a flange, to prevent its blowing through.

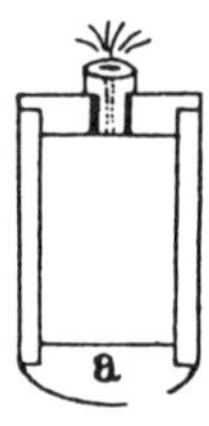
220

Another way of making shell and mine bottoms is to turn them between two centres, out of the grain end of the wood; for the shell bottom turn the outer end hemispherical, as at **a**, fig. 220. This gives additional strength.

Cylindrical shells, should, of course, be filled before the head is glued in; this can have the fuse previously fixed in, and ready.

ASTEROID ROCKETS.

97

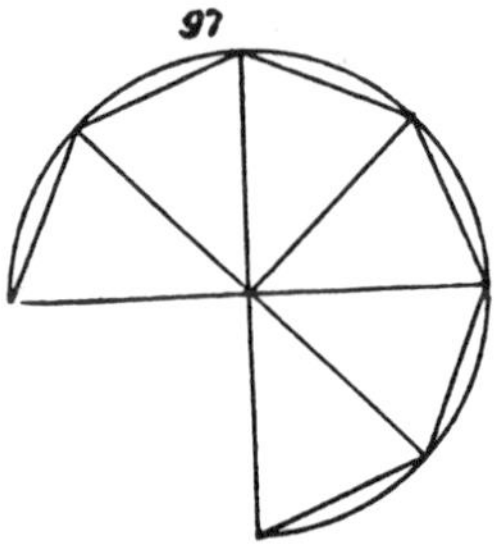

101

98

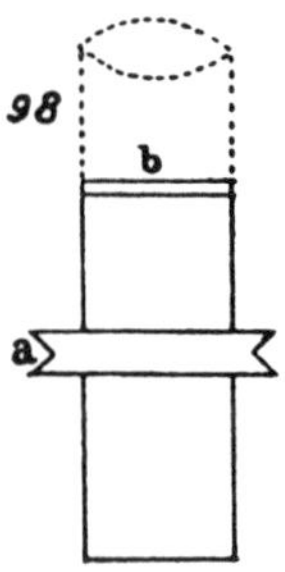

100

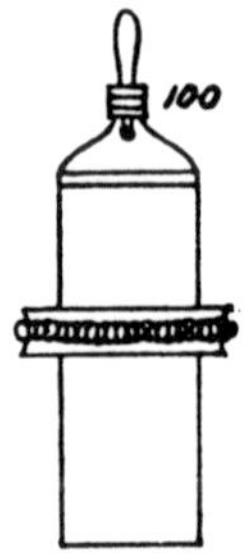

99

These differ from other rockets only in the head, which contains a parachute for floating a coloured case, attached to the strings. The parachute is best formed of silk, alpaca, or Indian muslin; it may also be made of glazed lining, or of tissue paper, carefully rubbed till the stiffness is removed. If of silk, alpaca, or Indian muslin, the parachute may be 30" square, with four strings attached to the corners; or it may be made of 6 triangular pieces, sewed together, like an umbrella. If of tissue paper, the paper may be 6/8 of a circle, or 6/8 of an octagon, as fig. 97. The tissue paper must be strengthened with a piece of crochet cotton, running round the edge, as shown by the dotted lines, fig. 101. It is simply laid on, and the paper snipped at each corner with the scissors, pasted and turned back. The radius of the circle may be 15".

For the lance, or colour, roll the case of three thicknesses of writing paper (about 10-1/2 diameters) on a 7/8" former: it may be 2½" or 3" long, and pasted all over. Turn a little wooden pulley, **a**, fig. 98, of a diameter to nearly fit the cylindrical head of the rocket; a hole in the centre of this pulley receives the colour; round the pulley, in the groove, lay a piece of chenille, fig. 100; twist the wires of the chenille 2 or 3 times together, and cut off flush. To charge the lance, set it on a flat surface, and drive in a film of mealpowder, or shell fuse; then 1/3" of red star composition; 1/3" of green, etc., till nearly full; stop the end with plaster of Paris, pressed in flat with a knife. Paste a strip of paper round the mouth, so as to overhang not more than 1/8"; lay in two short pieces of match, one across the other, and tuck in, or press down, the pasted overhanging edges; this will keep the match from dropping out. Glue the pulley on, about midway of the case; envelop the end **b** with a bit of glazed lining. Take a piece of string, fig. 99; bring the ends together, and tie in a knot. Insert the knot in the envelope, and tie round above it, as at fig. 100, leaving the loop out, to which to attach the strings of the parachute. Let the strings be 2 feet long: fasten them to the

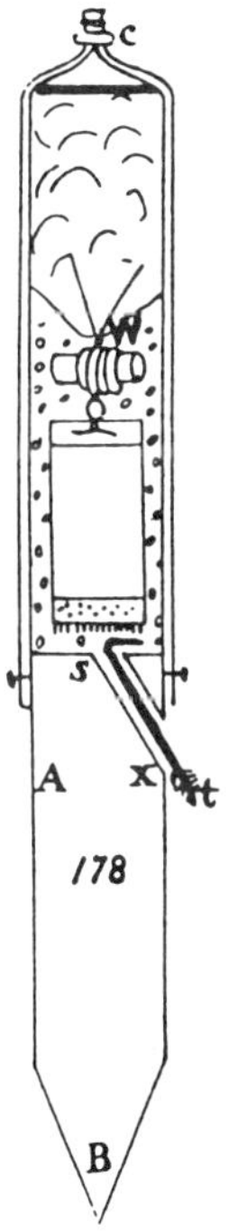

178

loop; gather them together; wind them round the colour, above the pulley; fold the parachute neatly, and bring the edges down over the strings, nearly to reach the pulley: this will keep the parachute a little open, and help it to expand. Both the tissue paper and the glazed lining parachutes I can vouch for; tissue paper I have never tried in a rocket, but am told it answers if it has not remained folded so long as to get set. I have experimented with it in the daytime, as in fig. 178, but could not get it to open.

Between the top of the rocket and the mouth of the colour is to be a little pad of grain powder. Make a little square or circular paper bag with tissue paper, and having put in the grain, and made the joining secure, smear both sides in the usual way, and sift dry meal over. If the rocket is to be fired at home, the parachute may be put into a long cylindrical paper bag, and left quite open at the top; but if intended to be carried to a distance, a thin bung may be put in at top, or a piece of turned wood; or a thin piece of paper may be pasted on it. In this case, be careful that it does not get stuck to the parachute.

The parachute and colour, with chenilled pulley, just described, are for the 8/8" winged rockets, hereafter mentioned. It was a plan used exclusively, I believe, by Mr. Southby: I have thought it worth preserving, but it is too tedious for general use. For smaller rockets a much less elaborate method may be employed, as follows.

177

For a 6/8" Rocket.

61

201

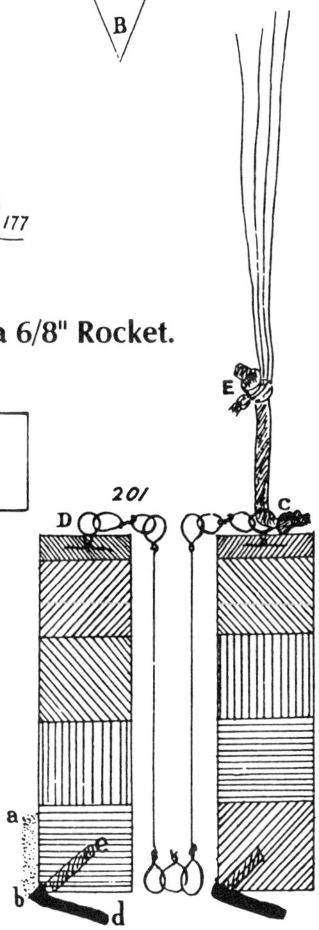

For the colour case take a piece of writing, cartridge, or 60 lb. kraft paper, 8" by 2¼"; paste it, and roll it on a 6/8" former, so as to make the case 2¼" long. Cut a notch in the lower end, as in fig. 61; bend an inch of match to the shape of the letter **V**, as directed for gold rains; set it half in, and half out of the case, through the notch, as **a**, **b**, **e**, fig. 201; put in a little lac-damped chlorate meal, on that a scoop of colour and mallet both together; proceed with other colours, till full within about 3/16": fill this void with thin plaster of Paris; and, before it sets, push in a wire of the size, and twisted to the shape of fig. 177, as shown at **c**, fig. 200. The match, **b a**, 201, will fall to the position **b d**. Or, instead of this way, the case may be set on a foot, fig. 9, turned with a tenon to enter 3/16"; fill with colour till nearly full, put in a bit of match, and secure it by pressing on it a little lac-damped chlorate meal, with the blade of a knife; remove the foot, and fix in the wire with plaster of Paris, as before.

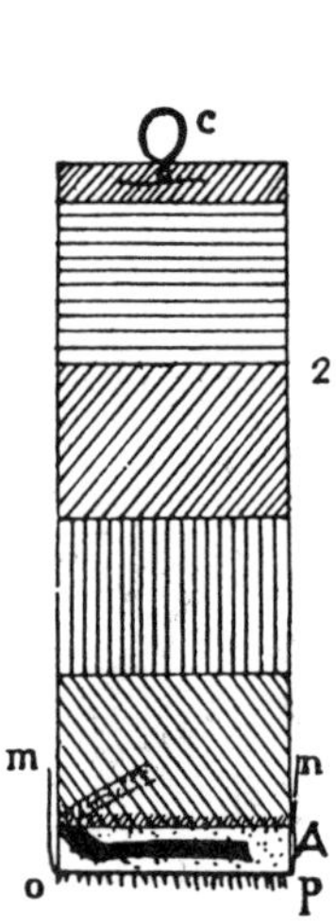

200

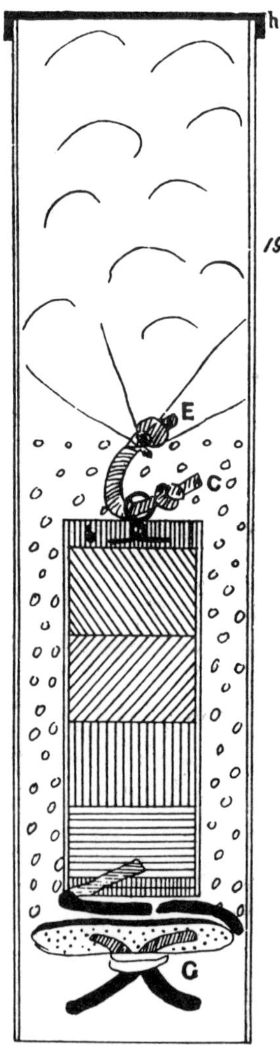

195

For a parachute take a piece of persian, 16" square, and tie to the corners 4 pieces of twine, about the thickness of **c d**, fig. 181, each 16" long; bring the 4 corners of the silk together; pull the strings straight, and tie them at the end in a knot. Now take a piece of laid cord, 1/8" thick, and 8" long, **E, C**, fig. 195 (drawn short, to suit the exigencies of the diagram), tie a knot at **C**, near one end; slip it through the ring, and tie a knot at the other end **E**; round the string, against this knot, tie the strings of the parachute.

The head of the rocket may be cylindrical, of 3 or 4 rounds of paper, pasted and rolled on an 11/8" or a 12/3" former, as fig. 195; or it may be of the shape and size of 196, the frustum of a cone **m o p n**, with a cone **o z p** attached. These may be formed with fig. 34 and 35; but it is possible to make them with a piece of cardboard for a former. Cut a piece of cardboard of the shape of **m o p n**, fig. 199; make **m n**, 1-3/4"; **o p**, 3¼"; **m o**, 4½". Spread out a sheet of kraft paper, lay the cardboard upon it, and mark round the edges, as **m o p n**, fig. 199; remove the card, and lay it at **n p s r**, and mark round, and so proceed till there are 6½" trapezoids. Cut out the piece with the scissors; paste it, turn it over, and lay it with the dry side upwards; lay the cardboard upon it, bend the paper over, press down, turn the card over, press again, and so on till the whole is folded; withdraw the card, and round the edge with the scissors. For the cone **o z p** describe 2 circles with a 2½" radius. Unite the cone to the frustum with pasted strips of tissue paper, about two thicknesses.

To fill the head, fig. 195, set it on a flat surface, fold the parachute, wind the strings round a cork, as shown at **w**, fig. 178, or coil them round the finger, as at **A**, fig. 196; push the parachute down the case, and press in upon it half an inch thickness of bran or pine saw-dust, firmly, to protect the thin strings. This being done, insert the colour case, and fill up all round with bran or sawdust till level with the mouth of the colour; if thought desirable a little cotton wadding may be pressed on the top, as at **a b**, fig. 196, to keep the bran in position; or a disc of cardboard, with a circular hole in it, may be employed for the same purpose.

Take a circular piece of touchpaper, put into it 16 grains of FFF, also a couple of pieces of match, and tie them in, as at **C**, fig. 195; or instead of this way, the FFF may be put into the mouth **A** of a touch-paper envelope **m o p n**, fig. 200; smear the surface **o p** with meal

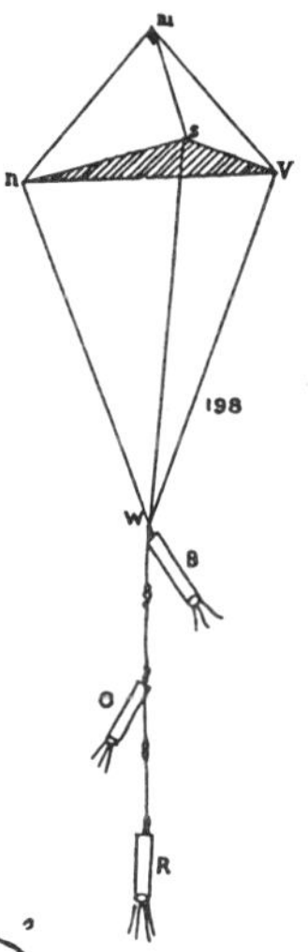

198

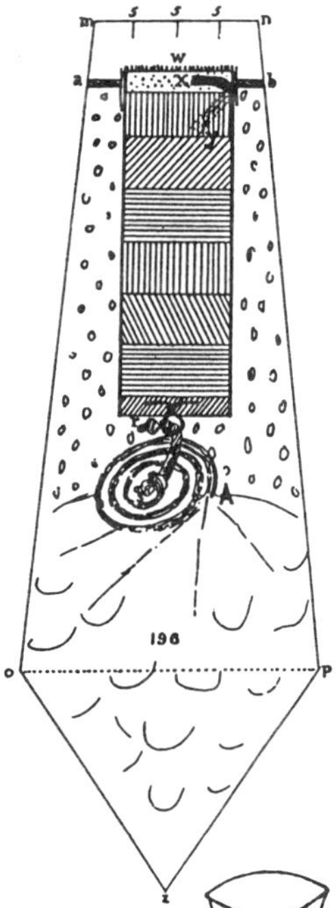

196

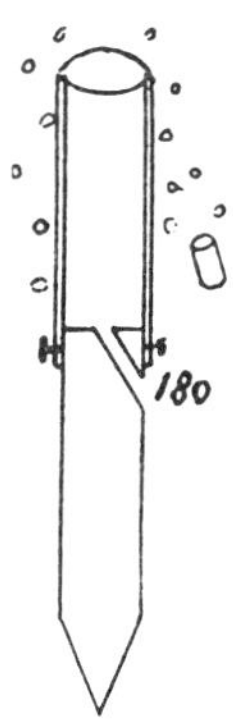

180

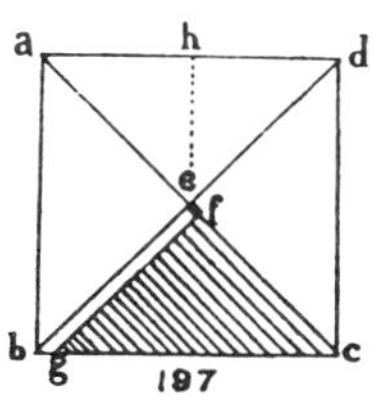

197

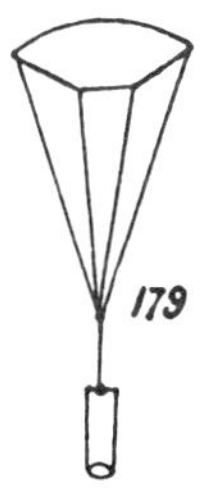

179

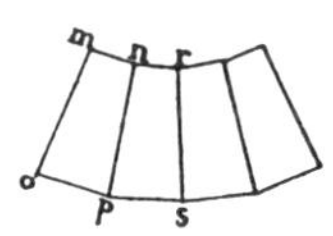

199

paste. If the head be of the shape of fig. 196, snip the head in the lines **s s s**, fit in the rocket, and secure it with a strip of pasted paper. If it be cylindrical, as fig. 195, paste a collar round the rocket, as **a b c d**, fig. 194. The cylindrical head may be closed at top by a pillbox-shaped lid, **f h**, fig. 195; or a thin bung may be fitted in; or a strip of tissue paper may be pasted at the edge, passed twice round, and tied as at C, fig. 178. The head, fig. 196, is to be filled upside down, in the same way as the cylindrical, which latter is drawn as it is to be fixed on the rocket. Another way of constructing a parachute is exhibited at fig. 198; take a piece of persian 20" square, **a b c d**, fig. 197; cut out the shaded part, **g f c**, leaving a narrow strip, **b e f g**; fold the upper portion, along the line **e h**, bringing the corner **d** upon **a**; paste the strip **b e f g**, and press it down; the parachute will then be a squat, or oblate, triangular pyramid, **n m v s**, fig. 198, with 3 strings.

To ascertain how a parachute will act the best way is to make the experiment in the daytime. Fill a case with sand instead of coloured fire. Take a piece of a broom handle, A B, fig. 178; point the bottom B; bore a sloping touch-hole, **s x**; insert a piece of match, touch-papered at the end, **t**; put the blowing charge at one end of the case and the parachute at the other; fit the rocket-head on, with 2 or 3 tin-tacks; set the post in the ground, and light the touchpaper. The parachute will be blown up several feet, as at fig. 179, and the bran will fall, as shown at the mouth of fig. 180. From this experiment it will be seen that the head itself should be strong with one weak part to give way; if it be cylindrical, let the weak part be **f h**, fig. 195; if fig. 196 be used, make it weak along the line **o p**. If the bran be not packed tight, or the quantity be insufficient, so that it falls loose, the flash from the bursting fire will reach the parachute, and instantly shrivel it into holes. To try how stars will light, fig. 180 may also be used; or they may be suspended from a nail in the wall, in a little triangular paper bag, with a piece of piped and touch-papered match protruding; on lighting they will fall, and it can be counted whether all catch.

TRIPLET ASTEROIDS.

Sometimes 2 or 3 parachutes are put, side by side, into a rocket head. Instead of this, three separate cases may be attached to one parachute. For this purpose, take some copper, or soft iron wire; bend 3 pieces to the shape and size of fig. 175; bend 2 more pieces to the shape of fig. 176; bend 4 more to the shape of A, fig. 174;

175

176

these should be as long as the colour-case, $2\frac{1}{4}$"; fig. 175, if preferred, may be first bent like fig. 145, then twisted across the middle into a figure of 8, as at **B**, fig. 174: link the 9 pieces together, as drawn. To each of the figure-of-eight links attach a colour, as at **B G R**, fig. 198; and at C and D, fig. 201; fasten the parachute at C. The links will now fold up and down, and the 3 colour-cases will lie side by side; put these into a cylindrical rocket head, in the usual way. Care must be taken that every ring is firmly twisted, so that it shall not come unfastened if it should happen to get red hot. The head, also, must be sufficiently large to hold the colour-cases loosely, that they may be blown out freely, and not entangle. Linked chains of greater length may be employed, if preferred, to keep the colours at greater distance from each other. The colours may be green, red, blue, so to remain; or the cases may be charged variously, by which arrangement the colours will appear to rise and fall; or, there may be at one time, 3 red, change to 3 blue, then to 3 green, as taste may dictate. The weight of the 3 cases should be the same as the one larger case which the parachute will buoy; for which purpose make the diameter of the former three 4/7 the diameter of the latter one; for, the length being the same, cylinders are to each other as the squares of their diameters. Now $4^2 = 16$, and $7^2 = 49$, where the latter square is about 3 times the former. If the single large case be 6 eighths, 6 x 4/7 - 24/7 = $3\frac{1}{2}$ eighths; so instead of one single case 6/8 diameter, put three cases, each of 7/16 diameter.

If a larger parachute were employed, it would support a great weight; but the parachute and its case are usually of less weight than the rocket will lift, that it may be carried to a greater height; and, in such cases, the stick requires to be cut a few inches shorter. It is not necessary to have a winged rocket.

HAND AND ILLUMINATION LIGHTS.

Hand Lights.

188

Cut a piece of 60 or 70 lb. kraft paper, 5" by 10"; paste it all over, and roll it on a brass-tube former, 1" diameter, so as to make the case 5" long. Have a turned piece of wood, fig. 188, consisting of a handle 4" long, and a tenon **a**, 1" long, and 1" diameter. Set the empty case on a flat surface, put in a pinch of dry mealpowder, and upon that a scoop of coloured fire; mallet it in, not too hard, till full within an inch; then glue it on the tenon, it will have the appearance of fig. 189. Cut two circular pieces of touch-paper, one 1½" diameter, one 2½". Paste the larger piece all over; lay the small piece flat on the composition; upon that the pasted piece; press it down smooth round the case. Cut a piece of white, or coloured tissue paper, 4" square; paste it, and wind it round the case.

Illumination Lights.

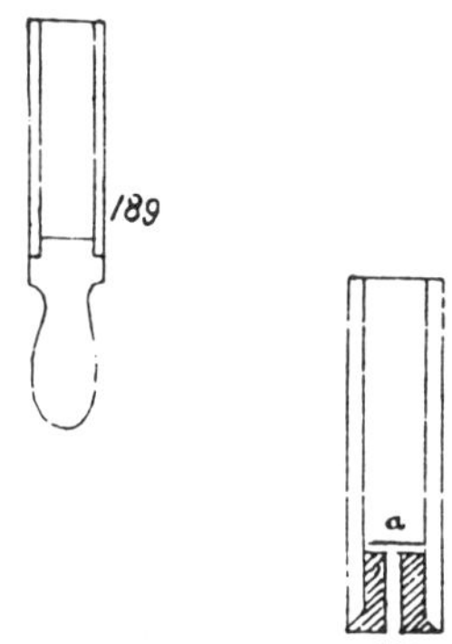

To make lights to hang to a post, take a cotton-reel, and saw it in two across the middle. Cut a strip of paper 4" or 5" wide, and long enough to go thrice round the half-reel; paste it and roll it round a former of the same diameter as the reel; when dry, glue the half-reel in one end, as at fig. 190, the flange outwards. To charge it, set the flange on a level surface; cut a piece of tissue paper, 3" or 4" square; paste it and roll it up into the shape of a marble; put it into the case, and drive it down flat with a solid drift, as at **a**, to form a bottom; then fill up with coloured fire, and finish like the hand-light. To fire it, drive a nail horizontally into a post and hang the light upon it, by the hole in the reel, fig. 191; the recrement, or scoria, will then fall, and leave the fire clear. Suitable compositions will be blue light, bengal light, wasp light; red, No. 23 or 24; green, 21 or 22; or any of the coloured-star mixtures, weakened by diminishing the potassium chlorate. Try a little, preparatively, in a pill-box. Should the brass-tube former not be exactly of the same diameter as the reel, it is best to have it a trifle larger, as the reel can have a strip of paper pasted round it, to make it a fit.

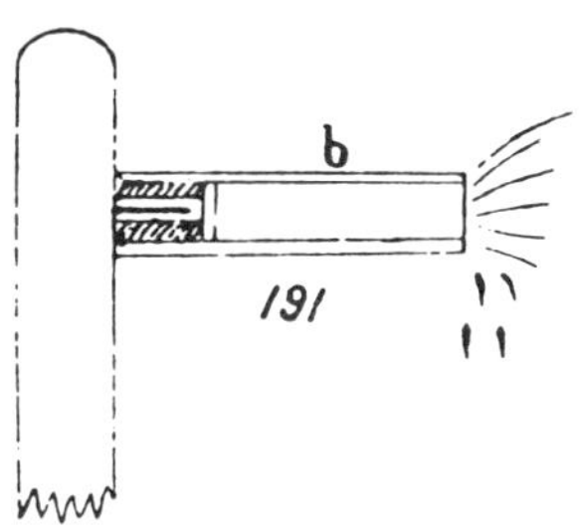

Hand and illumination lights, containing magnesium, may be made 3/8" diameter.

AQUATIC FIREWORKS.

Flying Fish.

Charge a 5/8" or 6/8" wheel case with any of the powders given under that name till full within 3/4". Turn a piece of light wood, pine, alder, or willow, fig. 237, into a cylinder 3" long, and equal in diameter to the exterior of the case; and with a tenon **a**, equal in diameter to the interior of the case, and 3/4" long, to fit into the void. Saw down the length, 2 kerfs or slits **xxxx, yyyy,** crossing each other at right angles. Glue the tenon in the case, or secure it with a couple of brads, **m, n,** fig. 239. Cut a piece of very thin wood, as the cedar of cigar boxes, fig. 238, into a parallelogram; make **ac** five diameters long, and two and a half wide, for a 5/8 case; that is

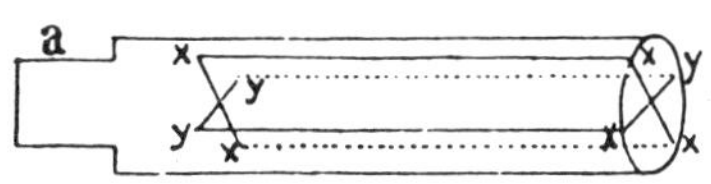

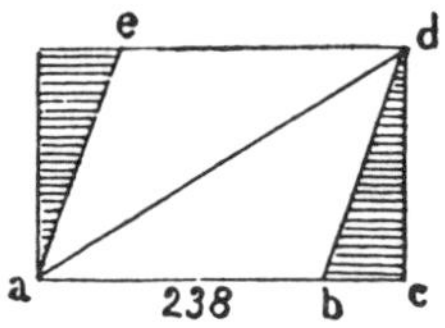
238

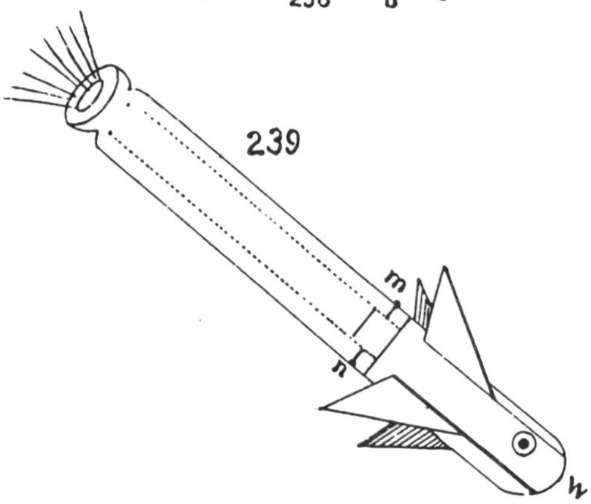
239

about 3" by $1\frac{1}{2}$"; mark from **b** to **c** one diameter, or 5/8"; draw the line **b d**, and parallel to it, **a e**; cut off the shaded portions; then saw along **a d**; this will furnish two scalene-triangular blades, or fins; make two more of the same size; glue the four into the kerfs, or longitudinal slits, putting the sides **ab, de**, inwards, and meeting in the middle. Round off the end at **w**, and stop up the kerfs with plaster of Paris. Paste a piece of tissue paper, or writing paper, round the case; give it two coats of size, or gum-water, allowing each to dry. Make a thick solution of shellac, in the proportion of 1 oz. of lac to 5 oz. of alcohol; varnish the case with this, with a camel-hair pencil, with two successive coats, for a protection from the water: or the case may be painted; or smeared with tallow, or covered with oiled paper, or with tinfoil. To fire it, hold it in the hand, light it, and throw it into a pond, lake, or river; the wood will cause it to swim, and it will dart along with extreme rapidity, at times rising out of the water and flying above the surface. The wood need not be varnished.

Dolphins, or Water-Devils.

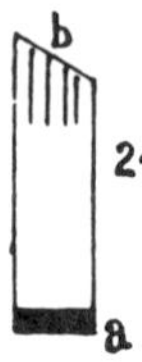
243

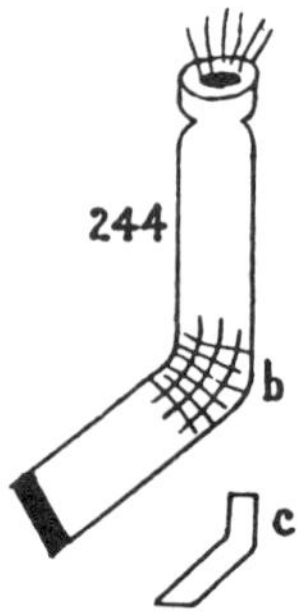
244

Charge a wheel case with No. 1 or No. 6. Make a thin case, fig. 243, of two or three thicknesses of writing or cartridge-paper, of a diameter to fit, on the wheel case, fig. 244. Plug the bottom **a**, with a layer of plaster of Paris, about $\frac{1}{4}$" thick; cut the other end, **b**, sloping, and slit it with the scissors, as drawn. Fit the slit end on the case, as at **b**, fig. 244; tie it firm, at an angle, with a piece of twine; cover with paper, size it, and varnish with lac-solution. Light it, and throw it into the water; the elbow acts like a rudder fixed anglewise in front of a vessel, and diverts it from a straight course, causing it to turn in all directions.

So small a case as a squib may be treated in the same way; the elbow-piece may be a bit of pine, or any light wood, having a tenon **c**, and sawn to form an angle.

Porpoises, or Water-Dragons.

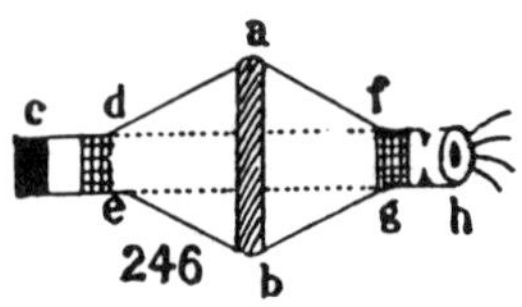
246

Take a case about 10 diameters long, as a 6/8" for example, $7\frac{1}{2}$" long; charge it with wheel-case powder till full within 3/4"; plug this with plaster of Paris, **C**, fig. 246, to make the end heavier, to dip. With a pair of compasses, stretched to a radius of 2/3 the length of the case, 5", describe a circle on a sheet of thick kraft paper; cut out 2 such circles, and paste them on one side; cut a straight line from the circumference to the centre, and wind the paper round

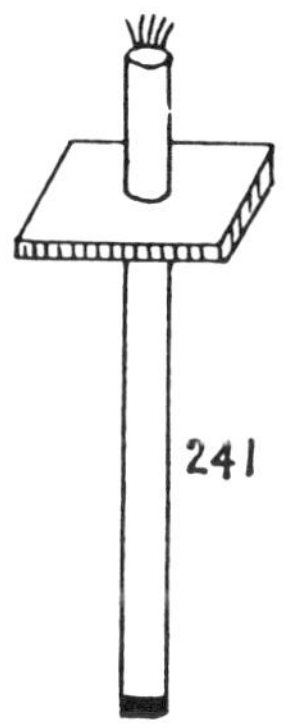

the fingers, pasted side outwards, into a cone. When dry cut off an inch of the point, making it like fig. 65, but with a much larger aperture. Slit the end thus cut, with the scissors, and fasten the 2 frustums on the case, as shown at fig. 246, tying them firm at **de** and **fg**. In order to fasten the large ends together, have a circular piece of wood, or a bung, **ab**, with a hole through it. This is to be, not in the middle of the case, but about an inch nearer **h** than **c**. Secure with pasted paper, and coat with a varnish of lac-solution. Instead of 2 paper cones, a wooden float, bored and turned in a lathe, like one on a fishing-line may be fitted on the case.

Floating Roman Candles and Gerbes.

Fix them in a collar of wood, as at fig. 241; a little sand may be put into the bottom, as ballast.

Jets, or Water-Fountains.

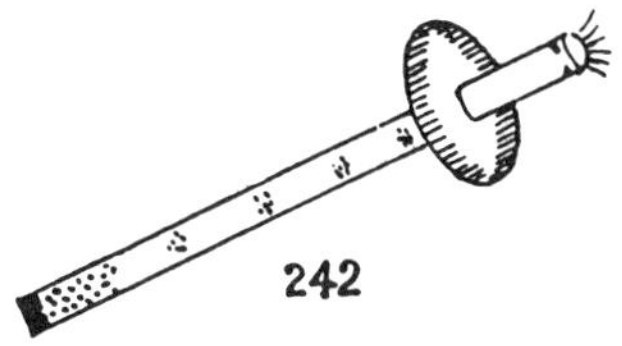

Charge a case, fig. 242, with wheel-case powder, No. 6 or No. 10; and, at intervals, put in a small scoop of grain powder, twice fig. 5, like a Roman candle without stars. Fit it with a collar of wood, of the shape of an onion. Light it, and throw it into the water. The explosions of the grainpowder will cause it to jump about, by fits and starts.

Spinners, or Water Wheels.

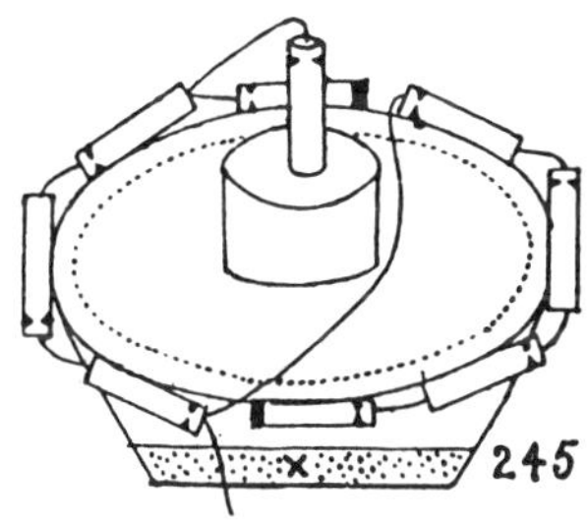

Cut a piece of pine plank into a hexagonal, or octagonal form, for 6 or 8 cases, and screw it on the top of a wooden bowl, fig. 245, having first put into it a quantity of sand, **x**, for ballast; the board must be a little larger than the bowl, so as to overhang all round at least ½". Bore holes in the plank, and tie round it 6 or 8 wheel cases, on the middle fix a mine. If there are 8 cases, let the first and fifth play together, which will balance the piece in its rotation. Put into the mouth of the first case a short port-fire, about ¼" diameter. Set the bowl in the water, light the port-fire, and push the bowl forward with a pole to a convenient distance. Should it be desired to pull it from the opposite side of a river into the middle, and to detach it to leave it free to rotate; pierce the port-fire in the part near the mouth of the case, run a piece of thin twine through it, and secure with a knot; tie a weight to the other end, throw it across, and as soon as the person is ready to pull it, light the port-fire; directly the twine is reached by the fire, it will, of course, be burnt in two, and can be drawn home, leaving the bowl unencumbered.

COMPOUND FIREWORKS.

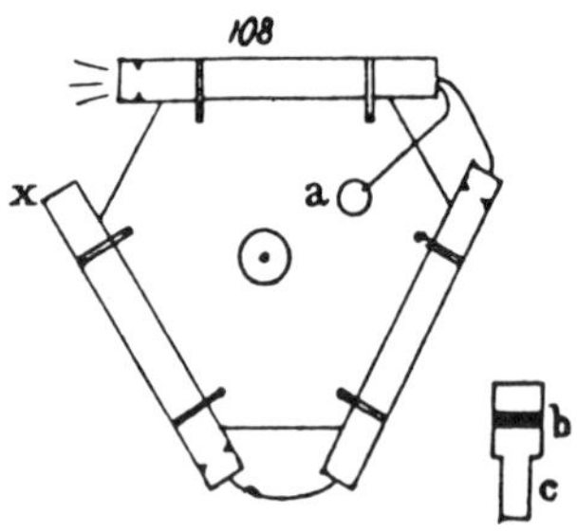

Fig. 108. A triangular: 3 wheel cases and a case colour, at **a**. These cases are tied either to a hexagonal piece of pine board, or to three spokes, radiating from a nave. The colour is sometimes tied to a nail, driven in to receive it; or it may be fixed on to a little peg. The peg is shown, by the side, at **b**. It is turned with a tenon, **c**; this tenon is glued in a hole, bored in the spoke; the part **b** may be ½" in length; through it bore a small hole; charge the case colour on a foot that enters the case half-an-inch; this void fits on to the peg, **b**; pierce the case, through the hole in the peg, with a bradawl; push a bit of binding wire through, and twist. If it be tied to a nail, or to a spoke, make it 1½" longer than necessary for the colour; charge this additional length with sawdust drawn firmly in; it can then be enveloped and laid crosswise upon a spoke, or straight on a nail, and tied, and the string covered with a gum-strip. Prime it with chlorate meal. The colour case may either stand at right angles, so as to face the spectator; or it may be fixed so as to lie in a plane with the wheel cases very slightly sloping upwards; the mouth of the colour case should point the same way as the mouths of the wheel cases; if it were placed in the opposite direction, it would meet the current of air, be blown back upon itself, and burn the case rapidly. With respect to the wheel cases, it is obvious that the mouths must all point in the same direction; the slightest consideration would suggest this; it is necessary, therefore, to have some certain rule of proceeding, so as to avoid mistakes. Now supposing, for the mere sake of illustration, if you had a case in your hand, it were to take fire, you would naturally wish the fire to be directed *from* you. Let this, then, be the guide. You are about to tie the cases on a wheel. Sit, to do so; take the wheel between your knees; place a case upon it, with the choke end *from* you; tie, near the choke, also near the end; turn the wheel, place another case upon it, choke end *from* you; and so forth. Without this certain way of proceeding, you would be very embarrassed with the 12 cases on fig. 112; but, by attending to it, not the slightest difficulty will be experienced. Some of the cases will point obliquely upwards, some downwards, but they will all point *from* you. In the end of the last case, as at **x**, fig. 108, put a little dry clay to prevent a stray spark igniting it, and tie the envelope; or tuck in the envelope, like as with an ounce of tobacco. Avoid every source of failure. Even the knots of the strings, with which the cases are tied on, are apt to come undone; they should either be touched with a dab of glue, or have a piece of paper pasted over them. Allow ½" of colour for every wheel case.

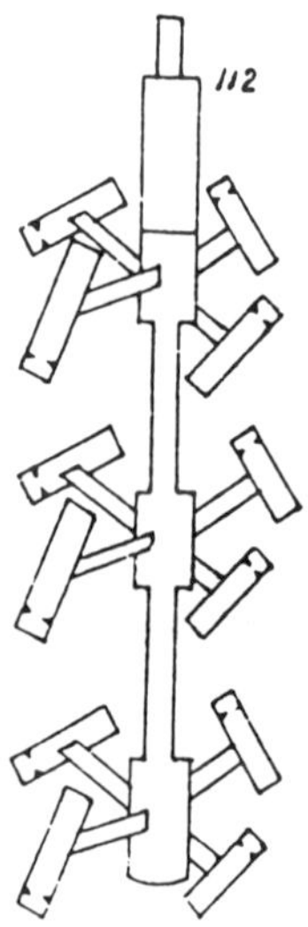

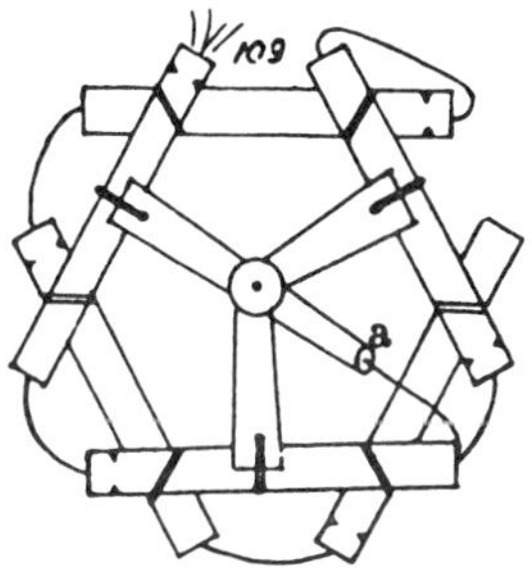

Fig. 109. A double triangular: 3 cases tied to the spokes; 3 to the previous 3. Double triangular frames are also made, with 6 spokes, on a long nave, 3 behind 3.

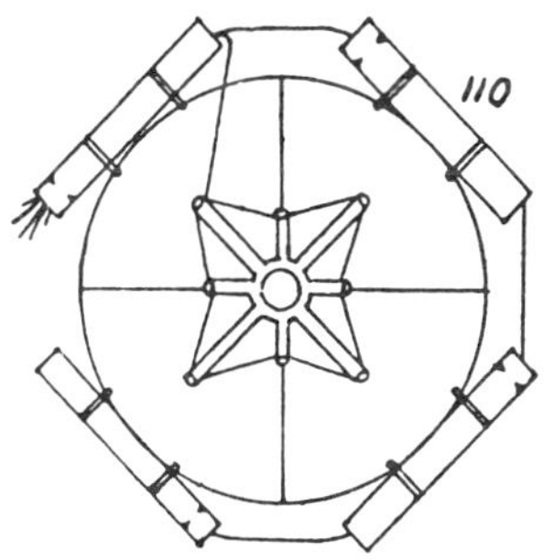

Fig. 110. A vertical wheel, illuminated. This is a wheel with spokes, and a rim, or felly. The wooden hoops of the toy shops will furnish the latter. They are fastened to the spokes with a nail, fig. 207. The illumination, as it is called, is made by lances, nailed with ½" Flemish tacks to the nave. The lances should be of different colours, but they ought, as nearly as possible, to keep time with each other. to effect this, it is necessary previously to adjust them, thus: Charge a number of cases, exactly 1" long, with different colours; use the same scoop to all, and give each the same number of blows, that they may be rammed as uniformly as possible. Fasten a leaden bullet to one end of a string, and tie a loop at the other end. Let the length from the top of the bullet to the top of the loop be 39". Suspend this from something, and set it swinging. Light the lances, one after another, and count the number of oscillations each endures. Keep a list of these, and write against them 5 seconds per inch, 6 seconds per inch. The bullet will indicate seconds, with the length of string recommended; and it matters not for so short a time, whether you give it a start of 6" or 12"; for if it goes twice as far it goes twice as fast, so that the swings isochronise; the inestimable discovery of Galileo, which led Huygens to the invention of the pendulum. In order that the lances may be nailed on, they should be charged upside down, and left with ½" vacancy; they can then be pinched flat, to receive the tack. Scrape out a little from the other end, and prime with very slightly lac-damped chlorate meal.

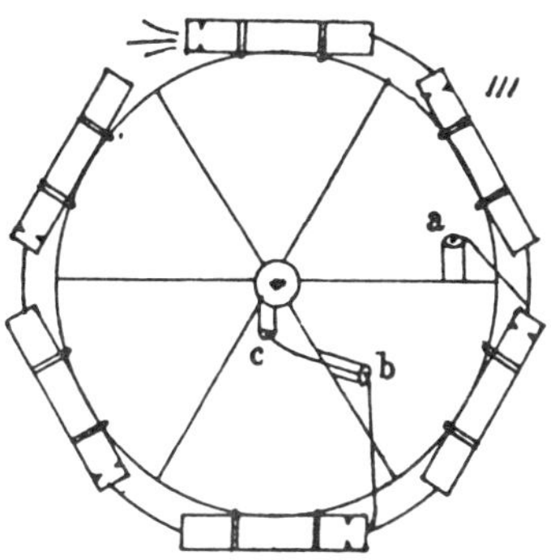

Fig. 111. A rainbow wheel. This is a vertical wheel, generally with 3 colours, as drawn; the tail of the second, or mouth of the third lights, **a**; the mouth of the fourth, **b** and **c**; but any arrangement may be made. Place the colours, red, green, blue, at different distances from the centre, so as to form rings, equidistant, when burning. Suppose the spoke 12" long: place the colours at 3, 6 and 9" from the centre. It receives its name from its resemblance to the rainbow.

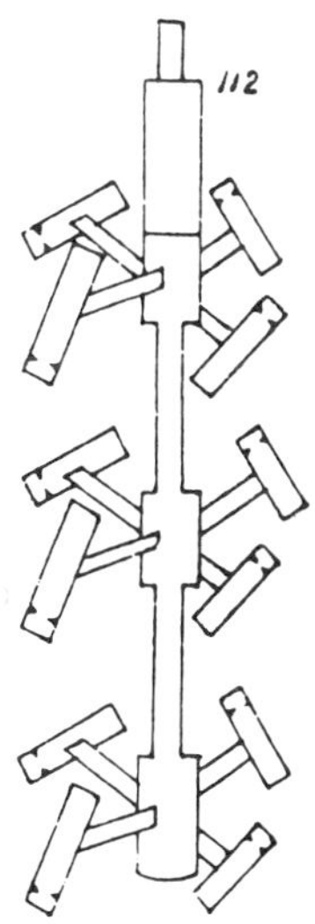

Fig. 112. A caprice, or furilona, according to the number of the cases. A caprice, from the capricious manner in which it turns up, down and round about, now this way, now that. A furilona, possibly, from the fury with which it plays, when 4 cases are burning

113

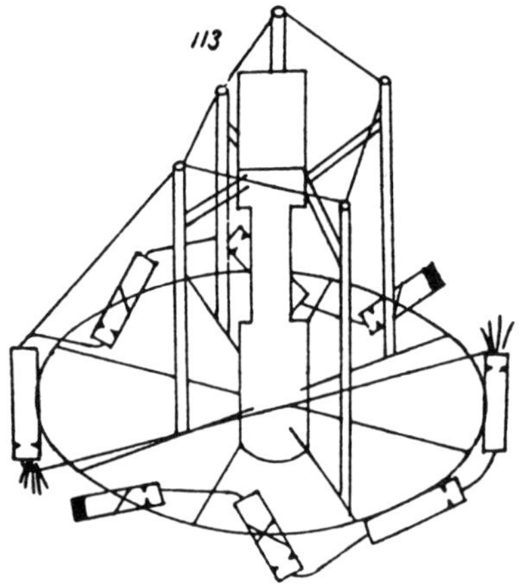

114

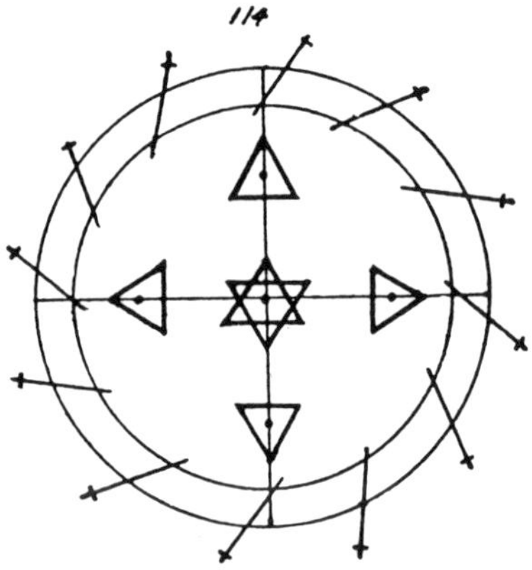

115

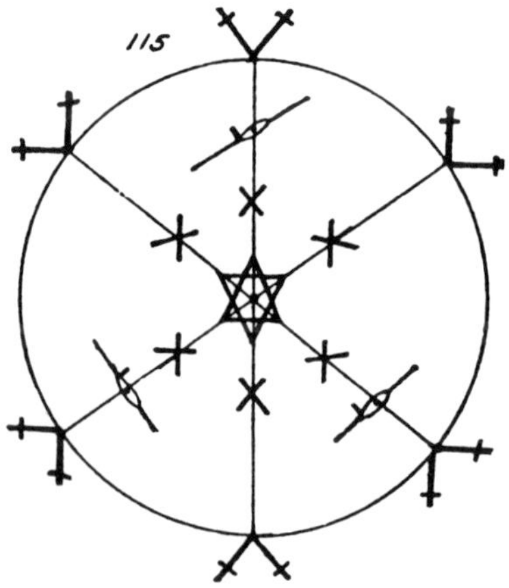

together at the end. A furilona and caprice wheel are much the same; the former generally has fewer cases on it than the latter. A coloured gerbe, placed on the top is very effective; or it may have a mine, or jack-in-the-box. The cases are to be placed so that some of them play horizontally, some obliquely upwards, some obliquely downwards; the spokes, which are concave at the end, are glued in, so as to determine the slope of the cases. If there are 10 cases, they may be fired thus, **h, ud, hud, hudp**; that is, 1 horizontal; 2 at once, 1 up, 1 down; 3 at once, horizontal, up, down; 4 at once, horizontal, up, down, perpendicular. If 13 cases, thus: **h, u, d, hu, hd, ud, hudp**. While tying the cases on the frame, it should be on a short wire, held perpendicularly in a vise, or block of wood; properly, the wire should be tapering, like a rocket spindle, so that the frame should bite when dropped on; the tapering will allow it to be turned round easily by slightly lifting it. Let the leaders be drawn straight, and not left dangling in curves, nor crossing each other. There must be *enough* match, but there ought to be no more. Look well to the mouths and tails of the cases; it is best to put a piece of pasted paper over each, for as the piece dashes round with great violence, if a stray spark falls on any compositon filtering out, the whole is spoilt.

Fig. 113. A horizontal wheel, with mine and Roman candles; the cases on the wheel are to be tied so that some play horizontally; some, obliquely upwards; some, obliquely downwards. To make a case play thus, tie it on the wheel across the middle of the case; this being done, turn the mouth upwards or downwards and tie it again, making the strings cross the previous ones, so as to form a letter **X**.

Fig. 114. A turning sun. Two concentric hoops nailed to spokes, or a frame; the cases lying on the hoops, slope; 3 or 4 may light at once; the spokes carry triangular, or vertical wheels; at the centre is a double triangular, or larger vertical.

Fig. 115. A rayonant star-piece. A wheel with six spokes; at the end of each spoke, two fixed cases, forming a **V**, the alternate spokes carrying saxons; at the centre, a double triangular; half-way between the triangles and the saxons, six five-pointed stars. A very beautiful piece.

116

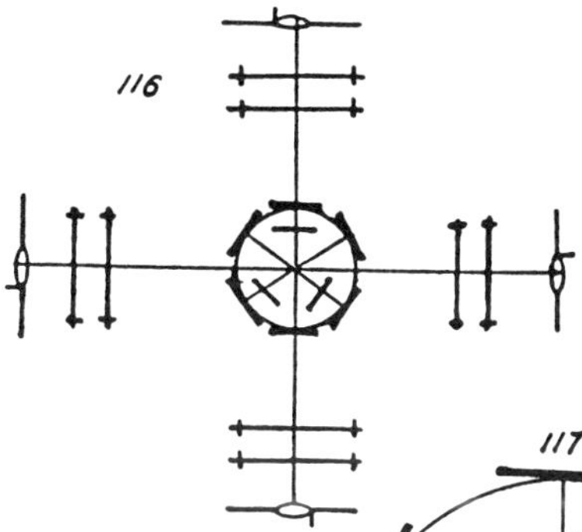

117

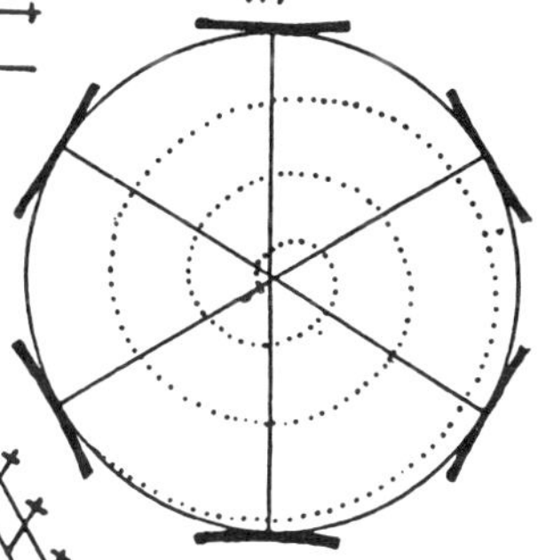

118

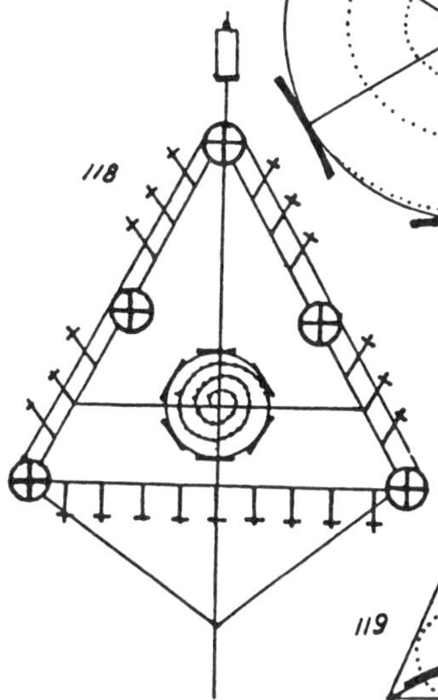

119

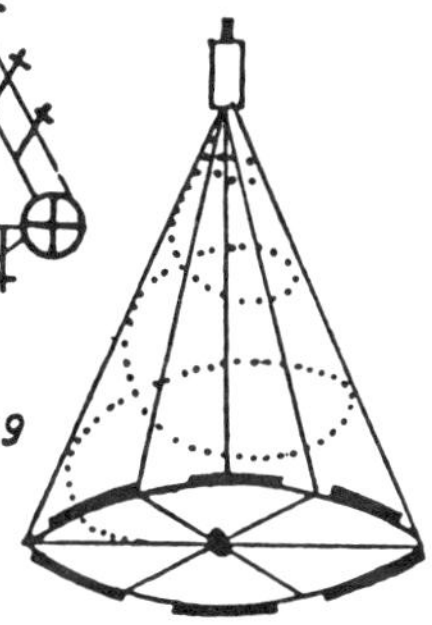

120

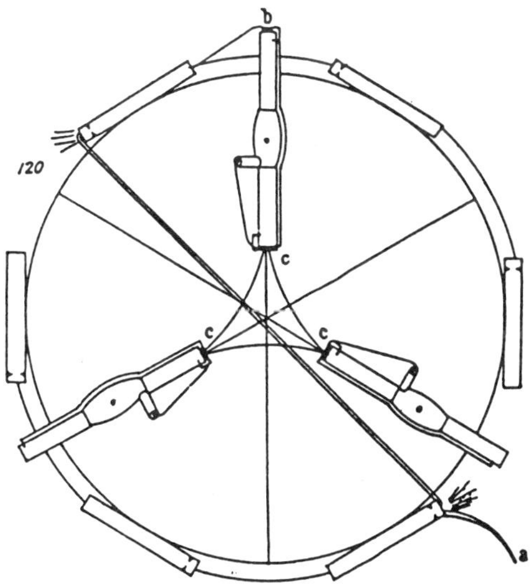

Fig. 116. A chequer-piece. A true-lover's-knot in the middle, 16 fixed cases, 4 on each side the true-lover's-knot; and 4 saxons at the extremities. The fires cross, and chequer into squares.

Fig. 117. A scroll wheel. Six or more cases on the wheel, to play in pairs; lances arranged on cane or hooping to form a scroll as indicated.

Fig. 118. A pyramidical piece. A scroll wheel in the middle; five horizontal wheels, or triangulars, at internvals, as represented by the rings; brilliant fixed cases playing obliquely upwards: at the bottom may be a row of cases playing downwards; these form what is called a cascade. Gerbes make the most effective cascades, but they require to be placed at a great height from the ground, if containing iron; the coke grains will be found suited for 8 or 10 feet.

Fig. 119. A spiral wheel. Six cases on a horizontal wheel; lances arranged in a spiral, on cane or hooping. It is shown more clearly at fig. 231; the lances stand upright, like balusters in a winding staircase; the leaders take the place of the handrail.

Fig. 120. A true-lover's-knot. Six 6/8" wheel cases, playing in pairs; three saxons, one carrying a blue, one a green, one a red colour. Light at **a**; this leader blows across, and lights the opposite starting case. The tail of this case lights the saxons: the ends of the saxons at **c, c, c,** before enveloping them, are to be smeared with meal; the end, **b**, is also to be smeared with wetted meal, to insure the ignition of the leader. This is a most beautiful piece: the colours, on the saxons, form loops, evolute, or involute, according as they turn left, or right; in the same direction as the driving-wheel, or the reverse, figs. 192 and 193; and represent, in a slight degree, the compound motions of planets and their satellites with regard to the sun. The centres of the saxons are carried round in a circle, like the earth in her orbit; the colours on the saxons revolve round the flying centres, like the moon round the earth. The wheel must not be less than 3 feet diameter; and it may be furnished with 8 wheel cases, if preferred.

Fig. 121. A revolving globe. This is also a most beautiful piece. The bottom is a horizontal wheel, carrying a strong half-hoop, **a, b,c**; a skeleton globe, formed with hoops, is suspended in this. This globe is driven by cases placed upon it round a hoop, crossing the other

121

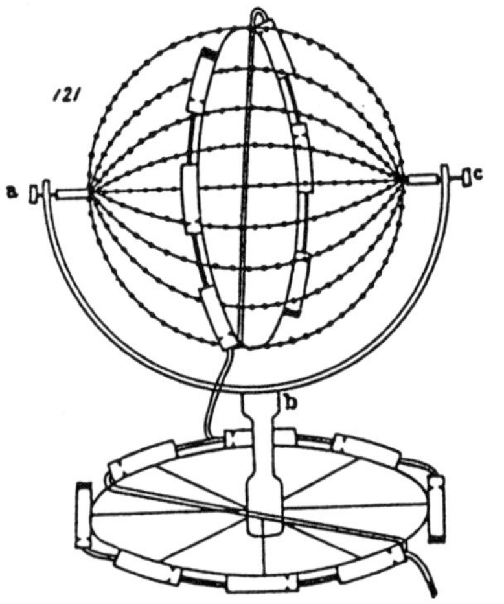

231

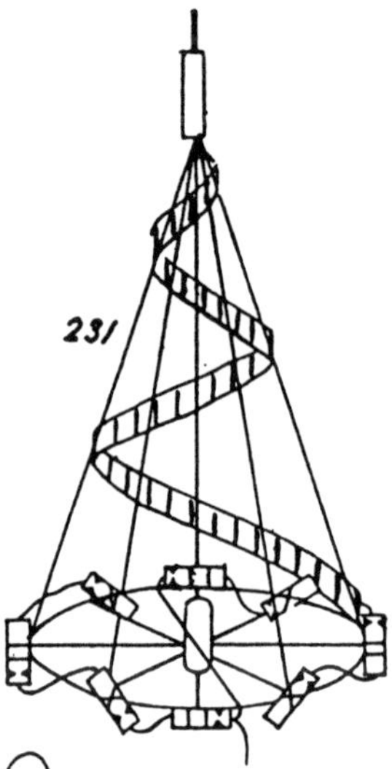

192

193

122

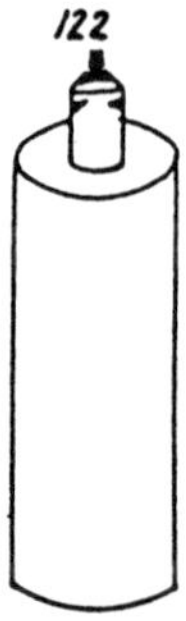

hoops at right angles, like the equator, at right angles to the meridians. The meridional hoops are covered with lances, white or coloured. The globe revolves vertically, while the wheel below turns it horizontally; the compound motion produces a peculiar oblique tumbling convolution, exceedingly perplexing to spectators ignorant of its construction. Instead of a globe, the top piece may be a revolving cylinder.

Fig. 122. A mine. This is a cylindrical case, 3 or 4 diameters high, containing serpents. The bottom of the mine should be a circular piece of wood, $\frac{1}{2}$" thick, glued in, or secured with 2 or 3 tacks. On it, place a circular bag, containing F grain powder. The bag is made with two circular pieces of tissue paper paper, one half-an-inch diameter larger than the other; lay the small one concentrically on the top of the large one; paste or gum the exposed rim of the bottom piece, and bend it up, and press it down on the small or top piece, almost all round, leaving a part through which to put the powder; when dry, put in the powder, and paste and bend down the part left, to stop up the hole. Put it into the mine, smear it with the brush dipped into meal paste, in the usual way, and with a pepper-box, shake in a little dry meal. Or it may be made like G, fig. 195, and placed match upwards. Take a fixed case, charged; envelop it so that the paper projects about an inch at the bottom; take a piece of squib-case, the same length as the serpents, put through it a piece of match, long enough to protrude at the top half-an-inch, and to bend over to form a hook; tie this in the envelope of the fixed case. Fill the mine with as many serpents, naked primed mouth downwards, as will stand upon the bag; with the scissors or a pair of pliers, draw out the middle serpent, put in the matched squib-case in its place; hold the fixed case upright in the mine, and ram pieces of torn paper tight round it, to offer resistance and cause the serpents to be blown higher. The easiest way of getting the bag into the mine is to have a cylindrical piece of wood, an inch or two longer than the mine, and about $\frac{1}{4}$" less in diameter; hold this upright with the left hand, lay the bag flat upon it, invert the mine with the right hand, and put it over the other, like an extinguisher; invert again, and withdraw the wood; the bag will be left in the bottom of the mine. Fifty shillings piled upon each other, may be got into a Roman candle case, by putting the case over them extinguisherwise.

To adjust the blowing-powder in the bag, use the following formula, I denoting the diameter in inches.

I x 2 I = drams.

Required the quantity of powder for a mine 1½" diameter.

1½ x 3 = 4½ drams = ¼ oz.
For 1-3/4"? 1-3/4 x 3½ =7/4 x 7/2 = 49/8 = 6 drams = 3/8 oz.
For 2"? 2 x 4 = 8 drams = ½ oz.
For 2½"? 2½ x 5 = 12½ drams = 3/4 oz.
For 3"? 3 x 6 = 18 drams = 1-1/8 oz.

It is by no means necessary to be exact; but the formula will serve as a guide.

123

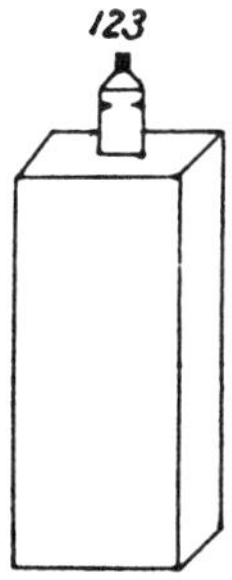

Fig. 123. A Jack-in-the-box. This is a case formed on a square prism of wood. The paper is to be pasted all over, but as it cannot be rolled it must be folded over, one side at a time, and rubbed smooth with a toothbrush handle, or pressed with the fingers. The bottom may be formed by bending in to a right angle, two opposite sides, and pressing them flat: then, upon them the two remaining sides, like packets of cocoa, etc. Or a square piece of wood, ½" thick, may be glued or nailed in for a bottom. Have the same charge of blowing-powder as for mines, only in a square bag; upon it lay a dozen or two of crackers matched like fig. 260, side by side, flat upon each other, sprinkle over them a scoop of mealpowder, set a flower-pot, or any short case on them, in the middle of the box, and ram in torn pieces of paper as with the mine. Cover with ornamental paper. As it is difficult to get the boxes always to dry perfectly square, should there be any little crevices in the bottom, they can be stopped with plaster of Paris.

124

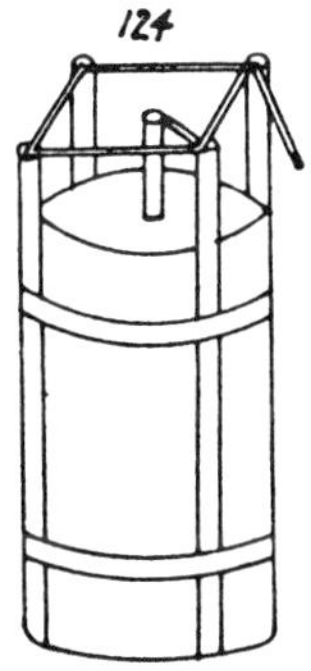

Fig. 124. A devil-among-the-tailors. A mixture of crackers and serpents, a Roman candle in the middle, having its bottom stopped with ¼" of shell fuse, instead of plaster of Paris to communicate the fire to the blowing-charge: 3, or 4, or any number of Roman candles, at regular distances, round the outside; these are to be tied with string, and a strip of pasted paper fastened round.

125

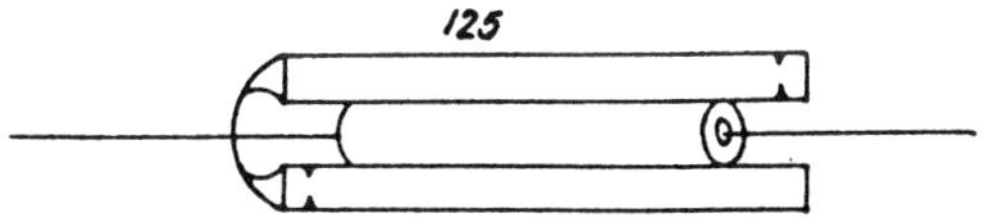

Fig. 125. A line or telegraph rocket. Two rockets tied to a piece of Roman candle case, head to tail: that is, one rocket tied in one

240

direction, the other in the reverse. This simply runs up and down a line. The rockets may also be tied like Fig. 240.

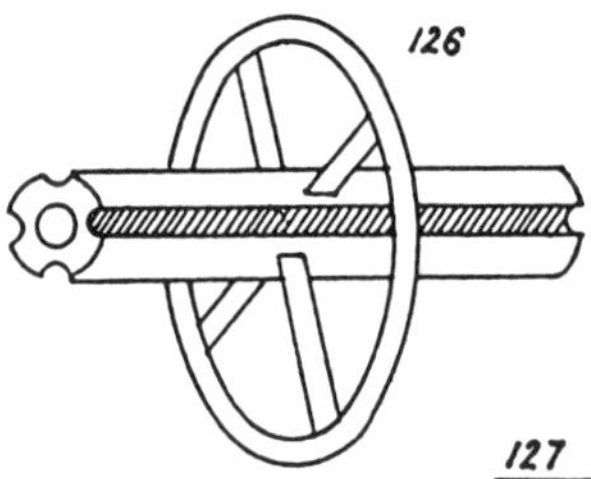

126

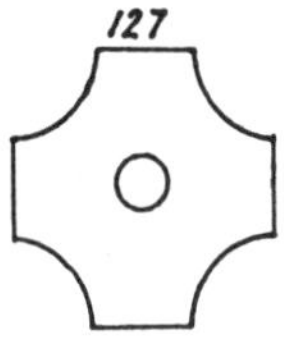

127

Fig. 126. A pigeon frame (of which fig. 127 is a cross section) may be made by taking a piece of pine, or alder, say 6" long and 1½" square; bore a hole through the length to receive the line upon which it is to run; plane off the corners, lengthwise, and channel them with a gouge, to form 4 semi-cylinders for the rockets to lie in. In the middle glue 4 spokes each about 3" long, and round the spokes nail a wheel. A piece of cane makes good spokes, light and strong. Tie on the frame, lying in the channels, with string passing round all four, four 6/8" rockets, two pointing one way, two the other, alternate; the clay, or plaster of Paris, on the top of the solid, in these rockets, is not to have a hole through it, as usual, but to remain perfectly closed. The piece is shown complete at Fig. 235. On the wheel, tie four 5/8" wheel cases, charged with wheel case powder, No. 1 or No. 10. Leader it thus: touch-paper the mouth of one wheel case. From the tail of this first wheel case, carry two leaders; one to the next wheel case, and one to the rocket whose mouth is nearest; the tail of the second wheel case lights the third wheel case and second rocket, and so on. A long clothes line must be stretched tight for this to run along. It should start 3 or 4 yards, at least, from the post to which the line is tied; and the rope should be so long that the first rocket could not carry it to the end; otherwise the leaders are apt to snap and spoil the piece. Towards the end, as the piece gets lighter, it will run the whole length of the rope without injury. It is necessary to indicate the first, or starting rocket, by some means; a piece of red paper may be pasted round the mouth, or an inch or two of string, or a button may be tied to it; or the match may be bent and tied so as to project an inch or two longer than the others. On fixing it on the line in the dark, you can then feel which is the starting rocket, and place its mouth *towards* the starting post without hesitation. Of course the pigeon runs away in the opposite direction. The best way of proceeding with the pigeon is first to tie one end of the line to the post, then slip on it a Roman candle case, then the pigeon, then another Roman candle case. Pull the line as tight as possible and tie it to the other post. Bring one Roman candle case against one post and push the other to the other post; this will prevent the piece from dashing against the posts and snapping the leaders. The length of the line must necessarily depend upon the extent of the premises; but from 60 to 90 feet will answer, and the height from the ground may be from 6

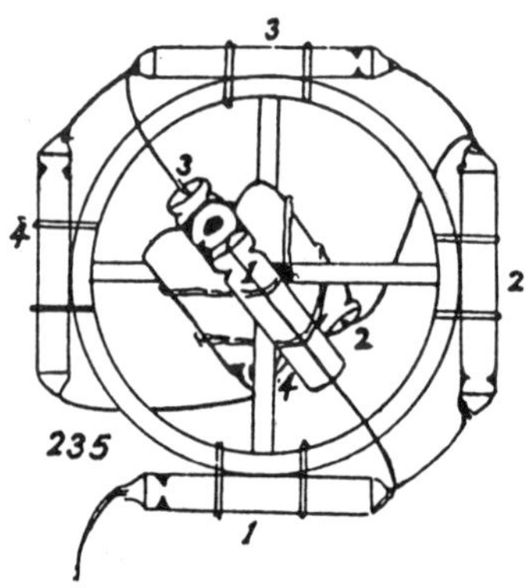

235

to 10 feet. Should the premises be very limited, the pigeon frame may be 4" long, the wheel 5" in diameter, the spokes, pieces of cane about as thick as a blacklead pencil, the hoop itself a piece of cane planed flat on each side: it may carry four 4/8 rockets and four 3/8 wheel cases. Cut the wheel case envelopes, then, $5\frac{1}{2}$" by 4" to go twice round the case, and overhang each end 3/4", the wheel cases themselves being only 4" long.

128

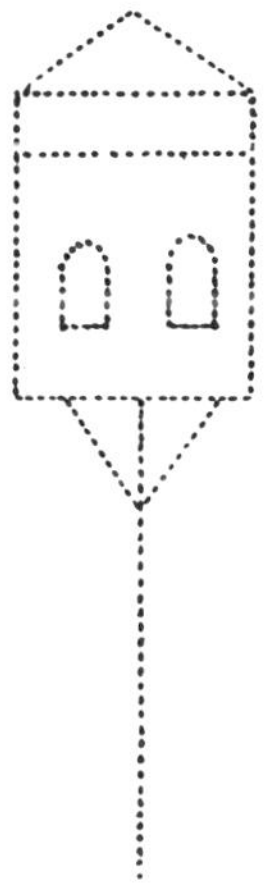

Fig. 128. A pigeon-house made with lances. This may be fixed in a central position and have 4 lines running from it: east, west, north and south. As soon as the lances are well alight, start the pigeons. The house may be simply a door, or pine board, about 2 feet wide, and $2\frac{1}{2}$ feet high, beside the triangular part at the top, the sloping sides of which may be 16"; or it may be a frame of the same size, made with strips of pine 1" square. The number of lances required for this size will be about 6 dozen for the house, and 2 dozen for the post and struts. If only one pigeon be used, it is not necessary to have a pigeon-house at all.

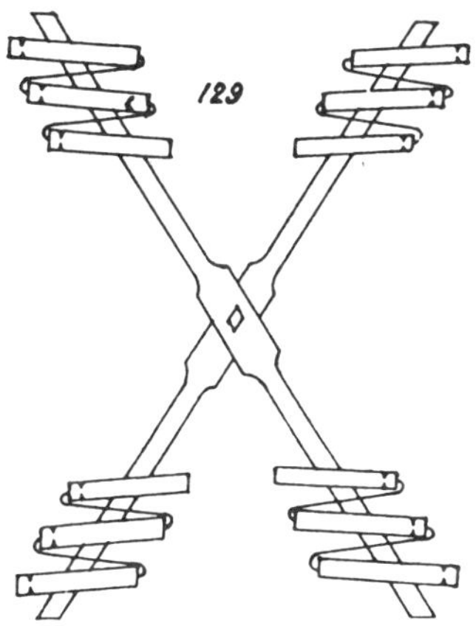

Fig. 129. A double guilloche, or windmill piece. This represents two windmills, turning in opposite directions; and imitates the engine-turning on the backs of watches.

Fig. 130. A five-pointed star. This has been already described.

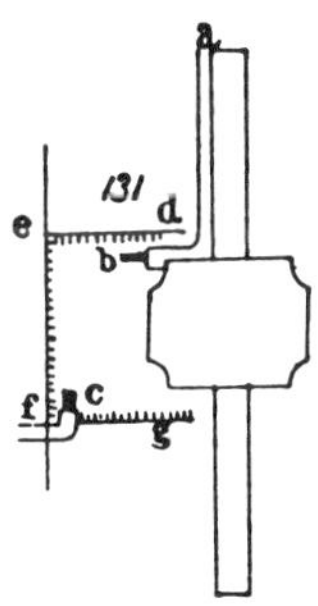

It is desirable, sometimes, to convey the fire from a movable or rotating piece, to a fixed or second rotating piece, which is effected thus. A leader, **a**, fig. 131, comes from the tail of the last case of the first rotating piece. The leader is tied, in one or more places, to one of the spokes of the wheel and to the nave, and is left protruding at **b**. A tin box, or a bit of a mine case, **d e f g**, is attached to the fixed or second movable piece in such a manner as to surround, but stand clear of, the end of the match, **b**, as it revolves with the wheel. A hole is made at **c**, and the end of a leader from the second piece brought through it. The inside surface of the box, **d e f g**, is smeared with mealpaste. When the leader catches at **a**, it blows through to **b**; this lights the smeared surface of the box; the flash communicates the fire to **c**, and this carries it where desired.

Sometimes a piece is made to drop, which is done thus: Suppose a cylindrical piece of iron, 6" long, 3" diameter, standing upright; on this a second cylinder, 12" long, 2" diameter; and on this a third

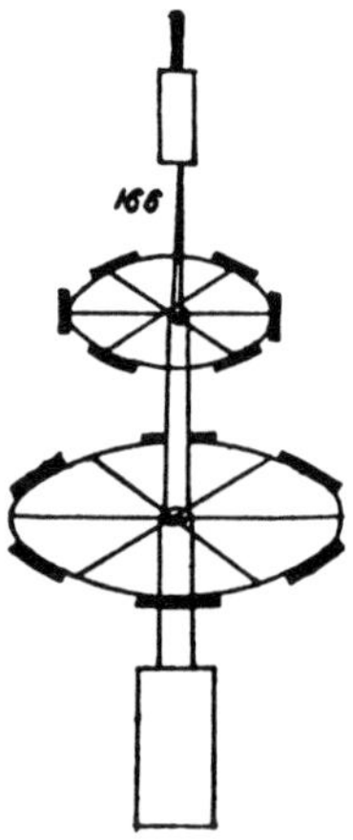

166

cylinder, 2" long, and 1" diameter. If now a brass ring $1\frac{1}{4}$" or $1\frac{1}{2}$" diameter be put on the top, it will slip down and rest on the 2" cylinder; if a ring $2\frac{1}{2}$" diameter be slipped over the top, it will fall and rest on the bottom 3" cylinder. It is obvious that if a horizontal wheel having a hole in the nave, $2\frac{1}{2}$" diameter be slipped on, it will drop and rest on the bottom cylinder; and that a second wheel with a $1\frac{1}{2}$" hole in the nave, will rest on the second cylinder. The two wheels then may be placed on and held together with a piece of thickly leadered match; the top wheel plays for a while and carries the second wheel with it; the tail of the first or other case lights the bottom wheel, and at the same time blows the match to pieces, and the wheel drops. Of course three or more wheels may drop one after another. See the design, fig. 166.

Sometimes a horizontal motion is changed to a vertical, thus: Suppose an upright wooden post, 3" square. Saw off the top 18", and fix it on again, with a hinge; drive a staple, on the other side, into the top piece; also one into the bottom piece; and connect the staples with a piece of string, to prevent the top piece falling. In the top drive a spindle, as usual, to receive a horizontal wheel. On the side of the fixed post, against the hinge, fasten an iron bracket, to stand forward, so that the top part, when the string is cut, can fall down and rest on the bracket. Now slip over the spindle a brass tube, 3" long, and almost a fit; then put on it a horizontal wheel, carrying a smeared box as above described, to convey the fire to a fixed piece; on the top of the spindle screw a nut. To the bottom post, tie a piece of shell-fuse, and let the string that prevents the top piece from falling be drawn over the mouth of this and secured. It is obvious that if, after the wheel has played awhile horizontally, the shell-fuse is lit by a leader, the string will be burnt in two, the top part will fall on to the bracket, and the horizontal motion will be changed to a vertical one.

254

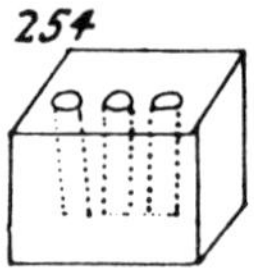

255

Roman-candles are best fired in bouquets of three or more; connect them with leaders about 3" long, fig. 255, and set them in a block of wood, containing three or more holes, diverging right and left, fig. 254; or tie them to struts of wood, nailed together, fig. 257, like the supports for garden flowers. Or three, or more, may be tied together, tight at the bottom; and at the top with a notched cork between them to make them strut.

257

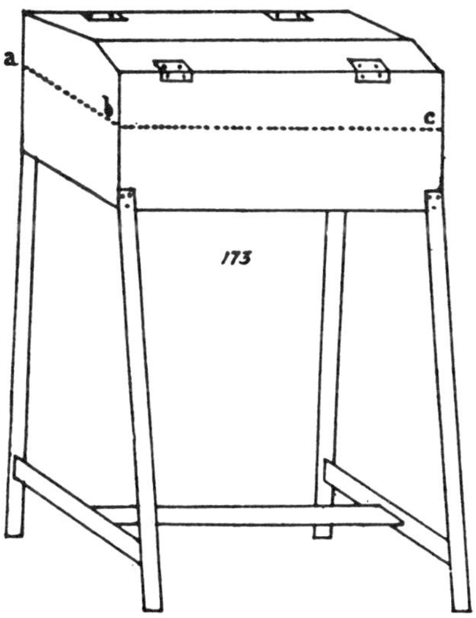

173

Two vertical wheels may be made to run round the outside of a table, to imitate the motion of the machine for grinding drugs, crushing clay, etc., like two parallel grindstones going round in a circular trough. Conceive a circular table, 20" diameter. With a gimlet bore a hole in the centre, large enough to admit a screw-eye, with liberty to turn easily round; slip through it a stair-rod, 2 feet long; it will overhang 2" at each end, and may be swung round. At the distance of about 4" from each end of the rod, imagine a cotton-reel, or a pulley; when the rod is moved round, the reels or pulleys will roll round in a circle, like the cylinder-crushers above mentioned. Now if a vertical wheel were fixed at each end of the rod, so as to hang outside the table, it is obvious that, upon firing it, the wheels would run round. Of course it must not be a screw-eye and a stair-rod; but anyone will understand from this description how to do it.

These contrivances are fallen somewhat into disuse since the introduction and variety of colours, but it is well to understand them.

To Fire a Girandole of 100 Rockets at Once.

172

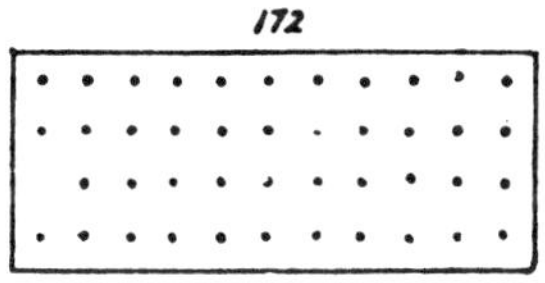

Suppose a cubical tea-chest. In the top, bore ten rows of holes, ten in a row, with a centre-bit; the same in the bottom, in such a way that the bottom holes fall perpendicularly under the top holes. Fasten the box upon four legs, one at each corner. Sift from a pepper-box a layer of meal powder over the top; put in the rockets, with their primed mouths, naked, to rest on the sifted meal. It is evident that, upon conveying fire to the meal by a leader, the flash will ignite the whole of the rockets at once. Of course it ought not to be a tea-chest, but a box constructed on purpose, with a penthouse lid, to fall over and protect the rockets till desired to be fired. It may be furnished with four legs, as in fig. 173. The shelf, fig. 172, fits in at **a b c**, fig. 173.

TOOLS.

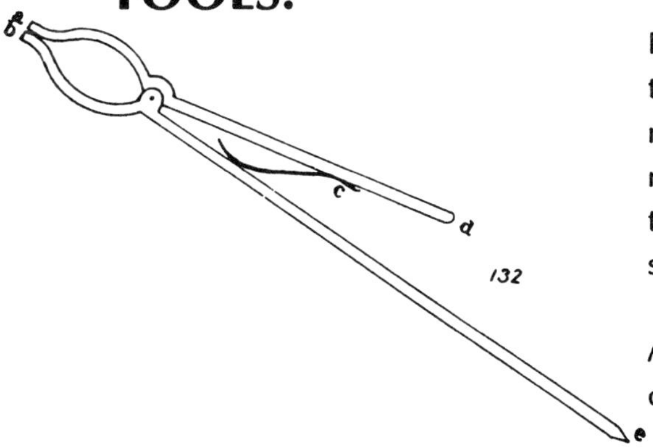

132

Fig. 132. A port-fire holder. This is made of steel, somewhat after the manner of a pair of pliers. The ends **a** and **b** are fluted, or channeled inside (semi-cylindrical) to receive the port-fire: a spring, riveted at **c**, holds it tight; by pressing on **d**, the burnt case drops; the end **e** is pointed to enter a long stick bored to receive it, and strengthened with a brass ferrule.

As an instrument of this kind can be obtained only when made to order, the following contrivance will serve the purpose of most

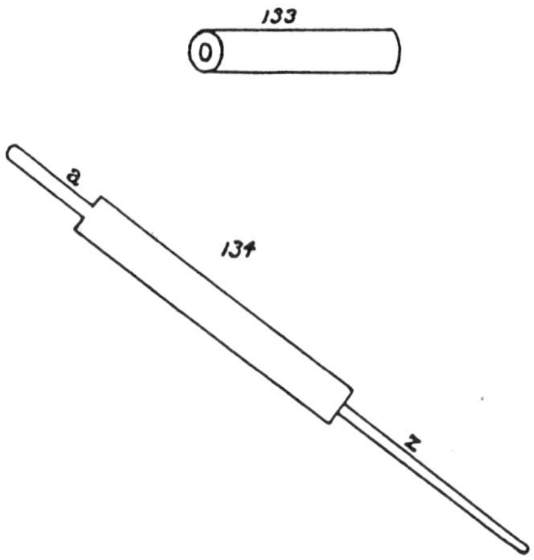

amateurs. First charge some little port-fires, 3" long and $\frac{1}{4}$" diameter, with the funnel and wire, till within about 1$\frac{1}{4}$" full; having filled a couple of dozen or so, invert them, and knock out the dust, as with squibs; then encircle the whole as if going to bang them; but pour in dry sand instead; empty a little sand from each, and stop the end with plaster of Paris. Scrape out a little compositon from the other end, and prime it with damped meal. Take 2" of a Roman candle case of a size to nearly fit the port-fire, or a piece of turned wood with a hole through it, fig. 133, 2" long. Have a long stick, fig. 134; cut the end, **a**, so that it will fit the hole in fig. 133, and enter half way up. At the other end fix a wire, **z**, 2$\frac{1}{2}$" long. By slipping fig. 133 on fig. 134 it is evident one inch vacancy is left to receive the sand end of the port-fire; when this is burnt out, it can be pushed out with the wire **z**.

TO CONSTRUCT A STEELYARD SCALE.

The numerous formulas for coloured fires render the use of very small scales excessively troublesome; and there is perpetual danger of losing the little weights for grains and pennyweights. By constructing a steelyard scale, the greatest facility is attained, one weight answering for all.

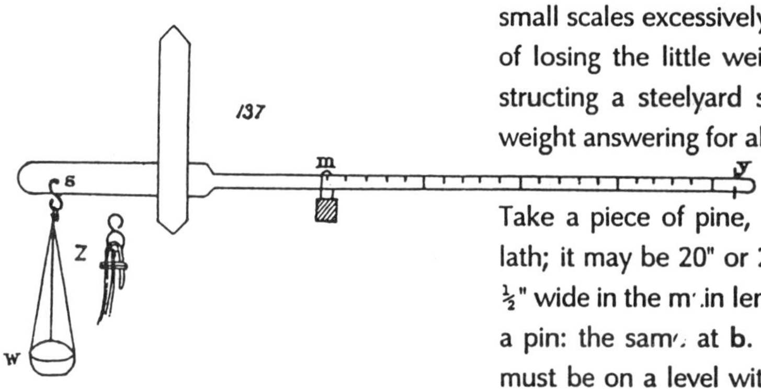

Take a piece of pine, fig. 135, about as thick as a venetian-blind lath; it may be 20" or 24" long, 1" wide, at the left end, and about $\frac{1}{2}$" wide in the main length. Make a hole at **a**, twice the diameter of a pin: the same at **b**. It is necessary to observe that the hole, **a**, must be on a level with the top of the long arm; if it should be a little higher, it would not be of any consequence: but it must on no account be lower. Have a piece of pine, fig. 136, about an inch square, and mortise a hole in it as drawn. Fix 135 in 136 so as to appear like 137. A bit of a common pin or needle is to go through the mortise and through the hole, **a**, of 135. See that it plays easily. At **s**, fig. 137, insert a piece of wire, bent like the figure 8; and to it, tie the strings fastened to **w**, which may be a copper bowl or a tin pattypan. The mode of fastening the threads to the figure of 8 wire, is simply to slip them through, bend them down, and tie round with thread as at **z**.

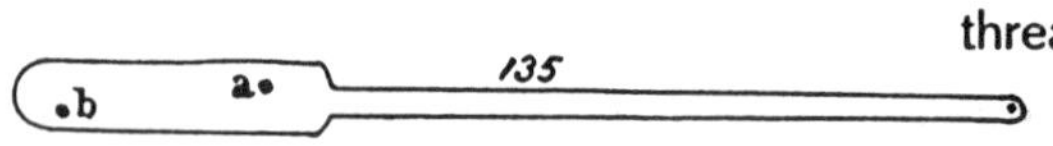

Take a piece of brass tube, an inch, or so long, fig. 138, and with some lead, melted in the bowl of a pipe, fix a bit of wire in it, bent like a staple, or the capital letter **U**, as drawn. This is for a weight. Slip it on the arm of fig. 137; put a wooden pin in the hole, **y**, to

171

keep the weight from falling off. Slip the weight along, till it balances the pattypan, **w**: suppose this point is at **m**. This will be the starting point. From **m** to **y** lay off a number of equal distances, as eighths of an inch. Number them and the scale is complete. If it be desired that 100 of these divisions should weigh 1 ounce, an ounce weight must be put in the pattypan, **w**, and the weight shifted along until it balances; the intervening space must then be divided into 100 equal parts. It is desirable to construct two; one with a large weight that will weigh 8 or 10 ounces; one with a small weight to weigh the 1/100th part of an ounce. This may be of wood with a wire staple about the thickness of a patent short white pin. Nothing can be more convenient than these steelyard scales, as one weight answers for all, and never drops and gets lost. It is well to have even a third steelyard scale. This may be made very thin, with a piece of wire, bent to the shape of fig. 171 for a weight. It may be made to weigh in its entire length only the fifth part of an ounce troy; so that if divided into 96 equal parts, it will weight the 1/480th part of an ounce or 1 grain troy; so that in trying a new colour, as little can be mixed as will charge only one or two pill-boxes. A little notch should be made at every division with a fine, triangular file, to prevent the weight from slipping: every tenth division can be numbered; this is best done with a blacklead pencil, as ink runs; and on the end, **s**, fig. 137, paste a piece of paper, and on it mark how many to the ounce. There is an instrument, somewhat similar, the chondrometer, for estimating the quality of grain; only this is constructed upon the principle of the log; the leverage and counterpose showing the weight of a pint, but indicating that of a bushel; as the half-minute sand-glass times the running out of the knots of the line, and indicates nautical miles, half a minute being 1/120 part of an hour, and the interval of the knots, 1/120 part of a nautical mile.

MONTGOLFIER BALLOONS.

To Project a Balloon.

139

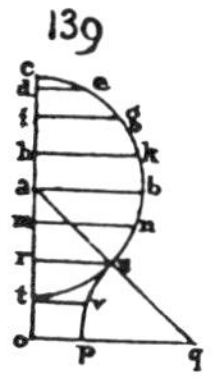

From the point **a**, fig. 139, with a radius, **a b**, of 5", describe a semicircle. Divide the upper quadrant into 4 equal parts, in the points **e g k**. Bisect the lower quadrant in the point **s**; draw the line, **a s**, and prolong it to **q**, making **s q** equal to the radius **a s**. From **q** with the same radius of 5", describe the reverse octant, **s p**. Bisect **b s** in **n**, and **s p** in **v**. Parallel to **a b** draw the ordinates **de, fg, hk, mn, rs, tv, op**.

To Form the Pattern Gore.

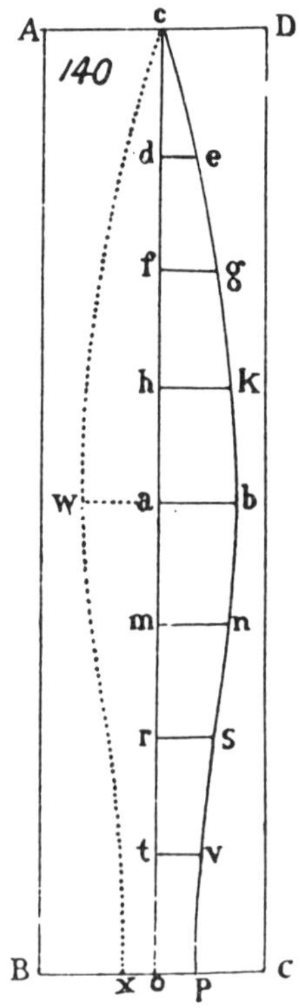

Take a piece of brown paper, such as used for laying under carpets, **A B C D**, fig. 140, 60" long and 12" wide. Fold it down the middle, **c o**, so as to make it double, 60" by 6". Press it flat, and divide the folded edge, **c o**, into 8 equal parts of 7½" each, in the points **d, f, h, a, m, r, t, o,** and draw straight lines perpendicular to the folded edge, and parallel to **AD** or **BC**. With a pair of compasses, or a slip of paper, measure the distances, **de**, fig. 139, and set it off from **d** to **e**, fig. 140; then **fg**, fig. 139, to **fg**, fig. 140; the same with all the others. Trace a curve line through the extremities, **e g k b n s v p,** cut through the double thickness with a pair of scissors, open it, lay it flat, and it will be of the shape drawn. Then cut it into two parts, along the line, **w a b**. The upper half is for the top part of the balloon; the lower half for the bottom. Instead of having a piece of paper 60" long, two pieces of 30" long and 12" wide may be used; one for the upper half and one for the lower. The radius of 5" is chosen to suit the size of tissue paper, which is 30" by 20", to get the gores as large as possible with the least waste; 60" are selected as the length of the gore, being twice the length of a sheet; 10" are 1/6 the length, and exactly half the breadth of the sheet, so the length and breadth are both employed without the slightest waste.

Instead of taking the trouble to construct fig. 139 with the compasses, the following measures will describe the pattern gore, fig. 140, at once: **de** will be 1-7/8"; **fg**, 3½"; **hk**, 4-5/8; **ab**, 5; **mn**, 4-5/8; **rs**, 3½; **tv**, 2-3/8; and **op**, 2-1/8".

There are to be 12 gores, consequently the equatorial circumference will be 120"; the diameter of the aperture 16". Four colours are recommended, yellow, green, blue and red, 3 sheets of each; but 6 of white and 6 of red have a pleasing effect. A smaller balloon may be made with six sheets, for which the radius will be 3½"; a larger balloon may be made with 24 sheets, for which the radius will be 7"; and a still larger with 48 sheets, for which the radius will be 10".

The upper half of the balloon, with the above measure, will be hemispherical, as shown at fig. 167. If the distances, **cd, df**, etc., be increased to 8 or 9", while the breadths of the gores remain unaltered, the balloon will be elongated, and become prolate-spheroidal or lemon-shaped: it is better, however, to elongate only the upper half; thus, while the four bottom distances remain 7½", the top four may be 8 or $8\frac{1}{2}$". If the distances, **cd**, **df**, etc., be diminished to 7 or 6", the balloon will be oblate-spheroidal, or onion-shaped, which is not an agreeable figure. The spherical form is the most elegant, and a 12-sheet is the most manageable size.

A balloon may be made of any shape by drawing a diagram, fig. 139, of the desired form, crossing it with ordinates and transferring the measures to a pattern gore.

To Cut the Gores.

141

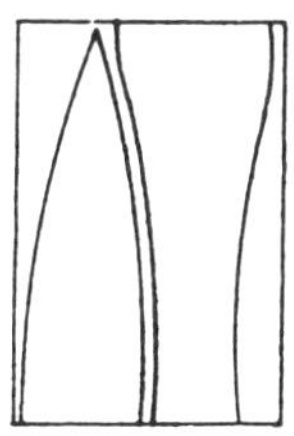

Lay the twelve sheets of tissue paper open and flat upon each other, as at fig. 141; place the two halves of the pattern-gore on the tissue paper; mark round them with a blacklead pencil; then with a strong pair of scissors cut through the whole twelve sheets at once. Practically the two halves of the pattern gore may be laid side by side, to touch each other, which will economise the cutting, though they are represented apart for distinctness.

To Join the Halves Together.

142

Lay the bottom part of the gore, fig. 142, flat upon the table, and the top part upon it, so as to leave $\frac{1}{4}$" along the edge exposed, as shown by the shaded part. Paste this, and without removing either piece, bend the pasted part of the lower half gore back upon the top, and press it smooth down. Gum may be used instead of paste, if preferred.

To Paste Two Gores Together.

143

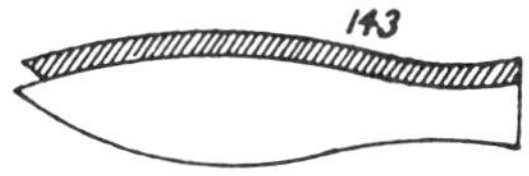

Lay one gore with the point towards the *left* hand, and another upon it, $\frac{1}{4}$" back, as in fig. 143. Paste the shaded part of the lower gore, bend it over and press as before. One of the gores, having been drawn in by pasting, is now narrower than the other. Finish the six pairs, and lay them by to dry.

To Paste the Pairs Together.

144

Lay one pair with the points towards the *right* hand, thus keeping the widest gore upwards; and another upon it, in the same manner, as fig. 144. Bend the top gore back upon itself, by folding it down the middle, as shown by the bottom shaded part, and lay a book or weight upon it to keep it out of the way; paste and double as before, and so proceed till the whole twelve are finished and lying upon each other. Be careful to disturb none of the gores. Laying them left and right causes every gore to be drawn in equally, and ensures regularity of shape. An additional neatness may be imparted to the balloon, by smoothing the joinings with a hot flat iron.

To Make the Last Joining.

Turn the points towards the *left* hand, lift up the top, and bend the ten inner gores back upon themselves; draw the top gore over so as to make it lie upon the bottom. Paste as before. This is a somewhat troublesome operation, and it is advisable to have the assistance of a second person.

To Wire the Balloon.

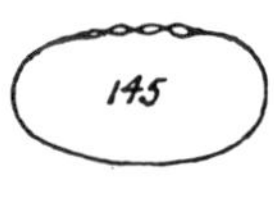

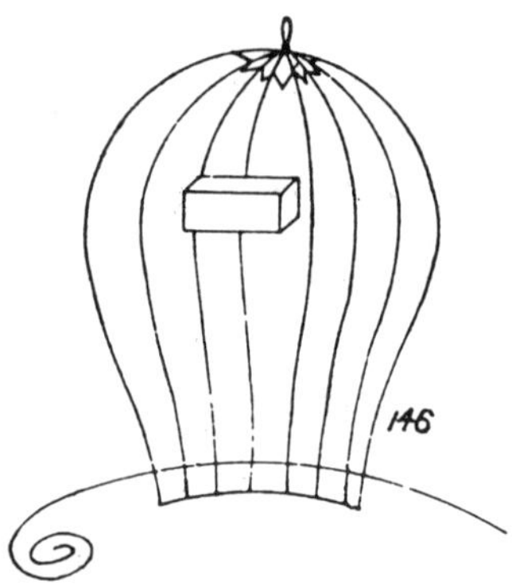

Select a coil of iron wire a little thicker than a pin, and remove the elasticity as recommended with coloured gerbes. Before applying it to the balloon, practise the method of making a joining, as shown by fig. 145, simply bringing the ends together, crossing them about an inch from the extremity, and winding them tight round each other. Now lay the balloon, as in fig. 146, with a book or a flat weight upon it to keep it down, curve the wire to the shape of the bottom of the balloon, and lay it about ½" from the edge. Notch the paper at the joinings of the gores, paste and fold back. When perfectly dry, not before, turn the balloon over, cut the wire 2" longer than to meet, bend the ends together and make the joining; paste and bend over the remaining half, and shape the wire to a circle. A star of tissue paper may be pasted on the top, and a thread passed through it to hold it by. To effect this open the balloon, pass a book up it, spread the top flat on the book, and paste on it a star or a circular piece of paper the size of a penny. If a hole gets torn in it, it can be repaired in the same way. In like manner a name may be put on it, as **VICTORIA, JUBILEE,** etc. Cut the letters out of 16 lb. tissue paper blue; let them be 10" high and 2" thick. Suppose the word to be **VICTORIA,** there are 8 letters; as they are to be placed in a circle, there will be 8 spaces. Having cut out the letters lay them in a straight line touching each other, the line will probably measure 72". The equatorial circumference is 120", then 120-72=48, and 48 divided by 8 = 6". Therefore paste the letters and stick them on, 6" apart. Rule a pencilled line 5" below the middle joining, on which to set them straight, they will then stand 5" below the joining and 5" above; the first letter may be 11" high. Of course if preferred, the balloon may be entirely white, but red and white are best when it is lettered. Take two pieces of wire, a little longer than the diameter of the mouth of the balloon, about 17"; pass

167

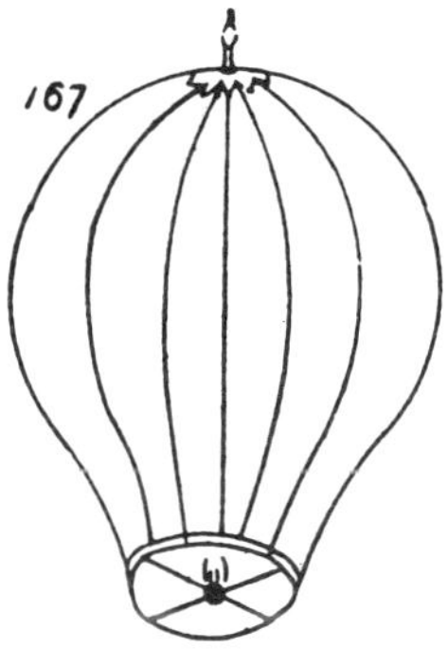

147

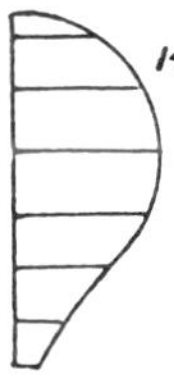

168

them half-way through a piece of sponge at right angles to each other; fasten these wires across the mouth; saturate the sponge with wood naphtha, or methylated alcohol, taking especial care that not a drop falls on the balloon; get someone to set fire to some pieces of paper on the ground; hold the balloon at a good height above the flame to prevent the alcohol in the sponge from catching. Three persons ought to assist in the start: one to manage the burning paper, two to hold the balloon steady, and to keep the mouth open for the ascending air to inflate it. As soon as the balloon gets expanded, set fire to the alcohol, and directly it begins to pull against the hand, which it will do, let it go. A balloon of this size will require an ounce of alcohol, and as much sponge as will absorb it--a piece rather larger than a walnut. See fig. 167. It should be started only in calm weather; it will not descend till long after the alcohol is consumed.

If the pattern gore be projected from fig. 147, the balloon will be pear-shaped, and may be filled with gas, like those started from the Crystal Palace. The paper should be 16 lb. tissue paper, and must have a coat of linseed oil. The bottom should be open, and fitted with a cylindrical neck, about 1" diameter, and 3" long, made of writing paper of three thicknesses. Fig. 168 shows the magnesium light attached.

A flat weight may be constructed by making a thin pine box, the size and shape of a matchbox; fill it with melted lead, nail a lid on and cover with pasted paper. It will also serve for a paper-weight.

A spindle, on which to fire wheels is shown at fig. 148; it is furnished with a wing nut, which can be taken off and put on without trouble. As there is considerable difficulty in obtaining an article of this kind, its place may be supplied by a carriage-bolt, of which hundreds, if necessary, may be procured of different thicknesses and of lengths varying from 1" to 12, or more. A thin one of 4" or 5" will be suitable for most purposes. As the nut is small and would be difficult to find if dropped on the ground on a dark night, proceed as follows. Turn a piece of beech or pine about $1\frac{1}{2}$" square and $2\frac{1}{2}$" long, into a cylinder with hemispherical ends, and a hole through it like a pop-gun. Cut it, while yet in the lathe, or saw it across in two pieces, one 1" long, the other $1\frac{1}{2}$". In the flat end of the inch length, mortise a hole and sink the nut in it: take a circular piece of wood, with a hole bored in the middle, and screw it flat on over the nut.

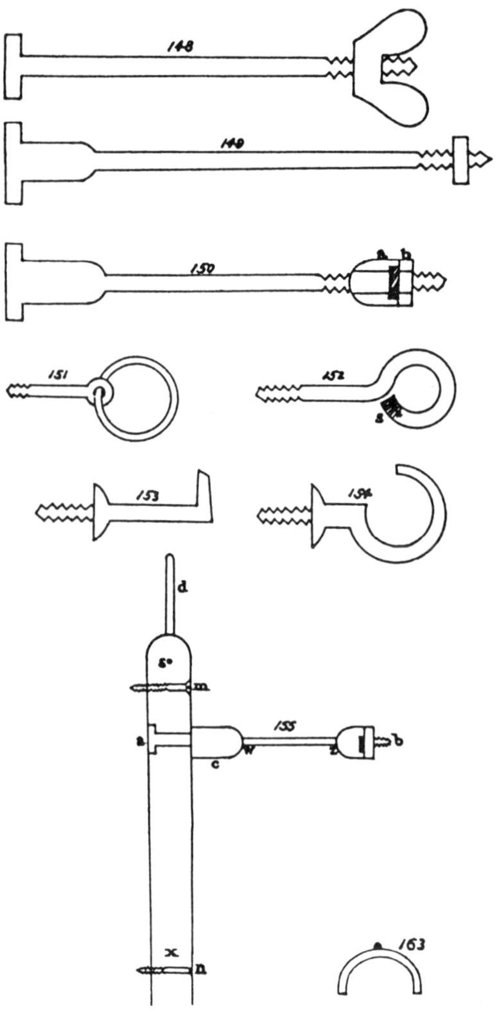

Fig. 150, **a**, shows the turned piece; **b**, the circular piece, screwed on to keep the nut from falling out; or instead of a piece of wood, a circular piece of tin or a brass washer may be fastened on with four tacks. This serves the purpose of the wing nut of fig. 148. The carriage-bolt is shown first at fig. 149, then at fig. 150, with the nut screwed on. Take a piece of pine, fig. 155, about 1½" square and 20" long: make a hole through at **a**, and drive the carriage-bolt from behind; the shoulder being square, will hold firm; in front, at **c**, slip on the above turned piece of wood, 1½" long: the piece with the nut embedded in it is, of course, put on at the end **b**. The hole through this inch-piece should be 1/8", or more, larger in diameter than the female (or socket, or counter) screw of the nut, in order that it may slip on and off with ease. At the top fix a strong wire, **d**, and file the end round and smooth: at **s** and **x**, in the side, fix two screw-eyes, or staples to receive rocket sticks. At **m** and **n** bore two holes, through which it can be screwed to a post or to the side of a set of steps. The screw-eye at **s** ought to be about level with the top of the post or steps. If there is too great a space between **w** and **z**, pieces of a Roman candle case may be slipped on; but a pop-gun, sawed into short lengths is better. Several should be kept in readiness of different lengths, to suit the naves of different wheels. A piece of iron rod, about ½" long, and to fit the hole, ought to be driven up furilonas, etc., to rest on the top of **d**, fig. 155; and the wire, **d**, should be a little longer than the hole, that the piece may play clear of the upright, the top of which should be hemispherical, as drawn. Fig. 151 is a screw-ring; 152, a screw-eye; 153, a screw-hook; 154, a cup-hook; 232, a ceiling-hook. A cotton-reel will furnish the end piece, **a b**, fig. 150, and the piece **c**, fig. 155, if the learner is unprovided with a lathe.

WINGED ROCKETS.

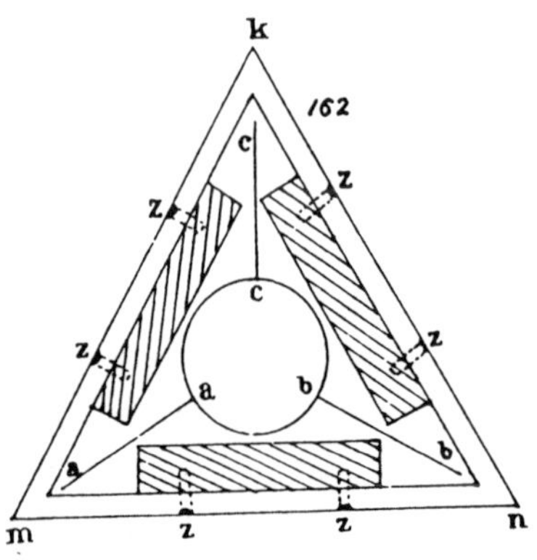

An 8/8" rocket is the smallest size worth making: it is, also, sufficiently large; and, as it requires a machine to fire it from, it is best to keep to one size. The head, fig. 156, may be the same as the external diameter of the case, or it may be slightly enlarged: to receive a parachute, it may be 12", or even 16" high; and the top may be closed with a lid, fig. 163, formed of a case choked tight, or by simply stopping it with a piece of turned wood, or with a thin bung pushed in as a cork into a bottle. A bung may be readily sawed thin with a fine-toothed saw; if required to be made smaller in diameter, it must be cut round sidewise, not longitudinally. If the head is to receive stars, it must not be above 6" high. To construct the three

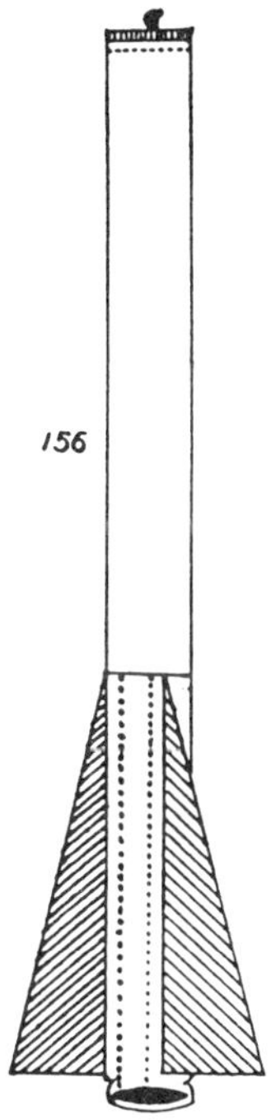

156

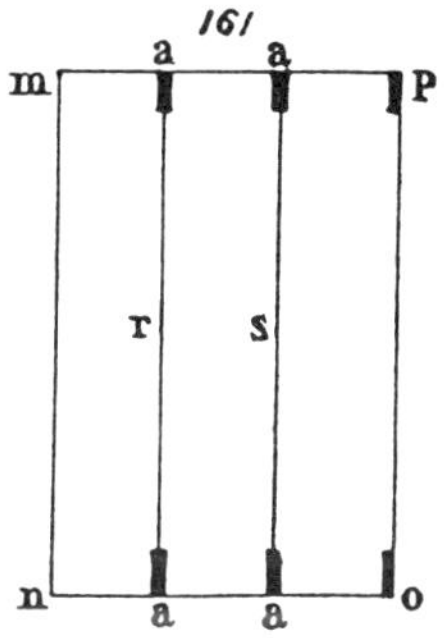

161

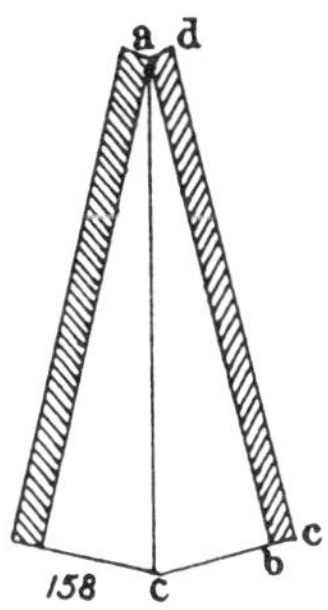

158

157

wings, take a thin piece of pine, or tin, and cut out a right-angled triangle, fig. 157. Make **a b** = $7\frac{1}{2}$", **b c** = $1\frac{1}{2}$". Take a piece of kraft paper, 8" long, 4" wide, and fold it down the middle lengthwise, so as to become 8" by 2". Lay the sloping edge, **a c**, of fig. 157 along the folded edge of the paper, **a c**, of fig. 158, and mark round it with a pencil. About 3/8" distant, draw the line **d e**, parallel to **a b**. Cut through with the scissors, and the paper will be of the shape of fig. 158. Bend the shaded parts flat, to form a flap, or kind of hinge. Paste the surface of the two triangles, and press them together to form a double thickness. If this does not make the wing stiff enough, insert, between them, a triangular piece of cardboard, to make it of three folds, like a shirt-collar. Take a piece of sheet tin, or a piece of cardboard, fig. 161. Let **m n** be 7", and **m p** such that it shall exactly wind round the rocket. Divide it into 3 equal parts, by the parallel lines, **r** and **s**, and at each end cut out little pieces, as shown at **a a**, etc. Now, if this be wound round the rocket, marks can be made through the slits, **a a**, with a blacklead pencil, or with a stencil-brush, dipped into blacklead powder, such as used for polishing grates. These marks will indicate the places for the three wings. Paste or glue them on, and secure the joinings with another slip of paper pasted over them, as shown by the dotted lines, besides the wings, fig. 156. The wings will branch out at angles of 120 degrees divergence from each other, like the lines **a a, b b, c c**, fig. 162. The rocket is thus winged; and, so far, complete.

To Construct a Slot-Tube, or Rocket Guide.

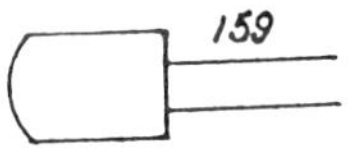

159

160

Procure 3 pieces of planed pine $\frac{1}{2}$" thick, 2" wide: let one piece be 6 feet 3" long; the other two, each 6 feet. Make 4 triangular iron holdfasts, **k m n**, fig. 162, 5" along each side; the iron may be a trifle above 1/8" thick, and 3/4" wide; in each side, let two holes be made, as in the usual holdfasts, to receive screws. The screws are represented at **z z**, fig. 162, which is a cross section of the tube. Fasten the 4 iron holdfasts round the 3 pieces of pine, as represented by **w w**, fig. 160, at equal distances. The pieces of wood will thus form a triangular spout with open corners. The pieces are to be level at the top; the longest one, of 6 ft. 3 in., will thus be 3" below the other two, as at **r r**, fig. 160. At **s s** bore two holes, about 1/8" diameter. Take a small piece of pine, fig. 159, and fix in it two pieces of wire, so as to have the appearance of a tuning-fork. These wires are to go through the holes, **s s**, of fig. 160, to form a resting-place, or support for the rocket, after it has been pushed up the tube. At the top of one of the short pieces, is to be a hook, **k**, by

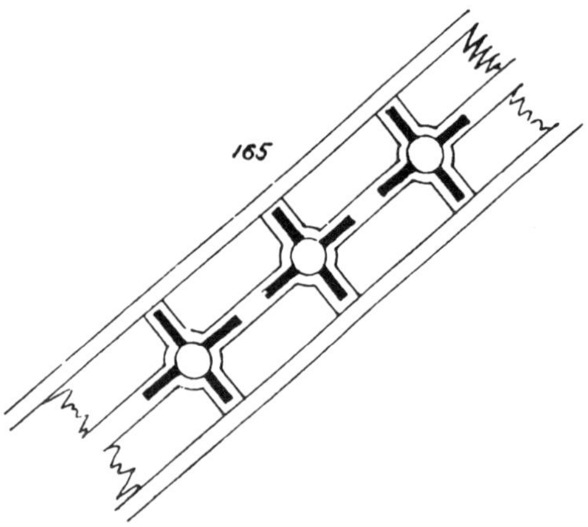

which to suspend it from a post. This post ought to be 9 or 10 feet high, that the rocket may be fired without having to stoop. Everything being ready, the rocket is to be put up through the bottom of the tube, and the wires of fig. 159 pushed through, for it to rest upon. It may then be fired.

These rockets are peculiarly fitted for asteroids, as, from their lightness, they rise to great heights, with immense rapidity, not in a continuous curve of uniform flexure, but with a sweeping serpentine motion, as indicated by the line, fig. 164. They might be employed, with great advantage, for distress rockets at sea.

They could also be made with four wings, and fired in volleys, from batteries constructed with square tubes. See the design, fig. 165.

METAL DRIFTS.

To Make a Set of 6/8 Metal Rocket Drifts.

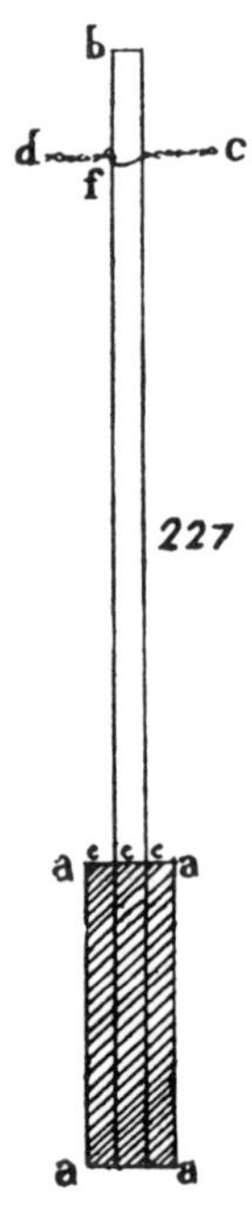

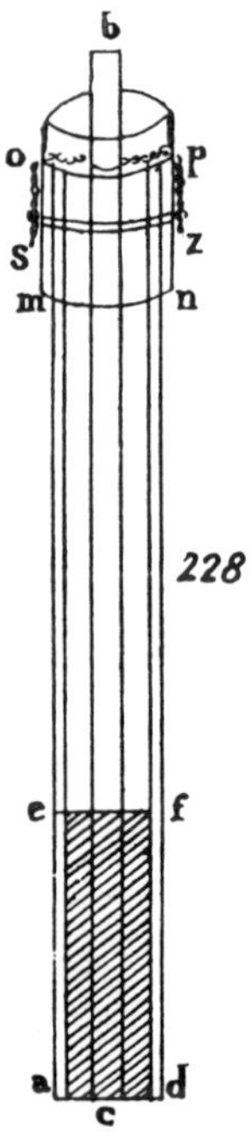

21

Procure about 13" of thick brass tube, 3/4" external diameter, saw or file it into 3 parts of the following lengths: 6¼", 4¼", and 2¼". See that they are perfectly dry and clean, and free from any oil that may have been used to facilitate the sawing. Take a straight piece of wire, 12" long, and as thick as the middle of the spindle, fig. 16, half-way between **b** and **c**. File it into 2 pieces, of 5" and 7". Cover the 7" length with a pasted piece of writing paper, and tighten it with the rolling-board, till it gauges the thickness of the bottom of the spindle, at **a**. When dry, cut a long strip of thick kraft paper, 1½", and wind it dry, as tight and true as possible, round the bottom of the covered 7" wire, **aaaa**, fig. 227, till it will exactly fit in the 6¼" tube; tie it round tight with a thread. Cut another strip, of the same breadth, and wind and secure it in like manner round the 5" wire, to fit in the 4¼" tube. Mix up a little blacklead, as used for grates, with a drop or two of gum-water, and, with a camel-hair pencil, paint the uncovered 5" wire, and the covered 7" wire with it; also the rolled edges of the brown paper, **ccc**. Now push the end, **b**, of the 7" wire up through the 6¼" tube, till the paper fits in perfectly tight, as at **aefd**, fig. 228, like the piston of a steam engine, or syringe. If the paper has been rolled true, the wire, **bc**, like the piston-rod, will now stand truly in the middle of the tube and perfectly upright, projecting at the top 3/4". Fix the 5" wire in the 4¼" tube in the same manner. Turn 3 pieces of hardwood, like fig. 21, with a head 1½" long, and a tenon 1½" long of a diameter to fit the inside of the tubes. Drive one into the 2¼" tube, which is to form the solid rammer. If the learner is unprovided with a lathe, he will be able, with a tenon-saw, a penknife, and a shoemaker's rasp, to make the heads out of a ruler, or a cricket stump. Set each of the three tubes upright in a jar, or pot, and pour in dry sand to keep them steady. Chop up 1½ lb. of lead into pieces with a hatchet; put into an iron ladle 6 ozs. of regulus of antimony; set the ladle in the middle of a strong clear fire; after the iron has become red-hot, the antimony will melt; as soon as it is melted, not before, put in the lead, also a piece of tallow, to prevent oxidation, and stir the metals together with a piece of wood. Remove the ladle from the fire, blow out the flame of the burning tallow, and pour the molten metal into each of the three tubes, till full. As soon as they are cold, fix the projecting end of the wire in a vise and pull the tube off, or drive the wire out with a hammer; the black-leaded surface will enable it to deliver. To get out the brown paper, insert a screw of suitable size, turning it round 2 or 3 times, and pull it out. The short solid drift will now be complete. Drive the tenons into the other 2 tubes; they will fill the void left by the paper plug. File the ends level and polish the tubes with fine glass paper or emery cloth; you will then have 3 heavy drifts, as beautifully true as if turned and bored in the lathe; one will be 7-3/4" long, one 5-3/4", and one 3-3/4". The tenons must fit tight, and reach the metal, or the metal will drive up till it reaches the tenon; should it do so, file the tube flush; it will not drive up again; but it is best to see that the tenon is the exact length before driving it in. Should the mouth of the holes be a trifle too small, they can be enlarged with a penknife, or the whole length, if necessary, can be bored larger with a shell-bit, or gimlet. Of course solid metal drifts can be prepared for charging wheel cases, saxons, tourbillions, etc.

The blackleading the rod prevents the metal from adhering: if a screw be blackleaded and supported in a bit of brass tube, and melted lead be poured round it, it can be unscrewed when cold; consequently, if the nut of fig. 149 should be lost, one might be cast by blackleading the screw, supporting it in an inch length of brass tube, and pouring in melted lead; it would serve at once, instead of the prepared nut, **ab**, fig. 150. If a screw be smeared with oil or tallow, and plaster of Paris poured round it and allowed to dry, it can be unscrewed easily; the tallow or oil prevents the plaster from adhering, as blacklead does the metal; tallowed screws do not get set, and blackleaded surfaces cannot be soldered.

The term blacklead, the English translation of the Latin plumbago, is a misnomer given in times of ignorance, and leads to the idea that it is a metal; whereas 96 per cent, or 24/25 of it is carbon; the remaining 1/25 consists of iron, silica and titanium; a more appropriate name is **graphite**; it has no connection whatever with lead. The two hollow drifts just described and the one solid will be quite sufficient; and the better to insure this, taper the spindle no more than will just make it deliver, see fig. 194. These drifts, from their weight, fall with great force upon the composition, and consolidate it most effectively. It is advisable to set the pot out of doors; then if any of the melted metal gets spilt it can be picked up when cold; whereas, if the operation is carried on by the fireside, and any of the metal falls upon the hearthrug, it will be next to impossible to remove it.

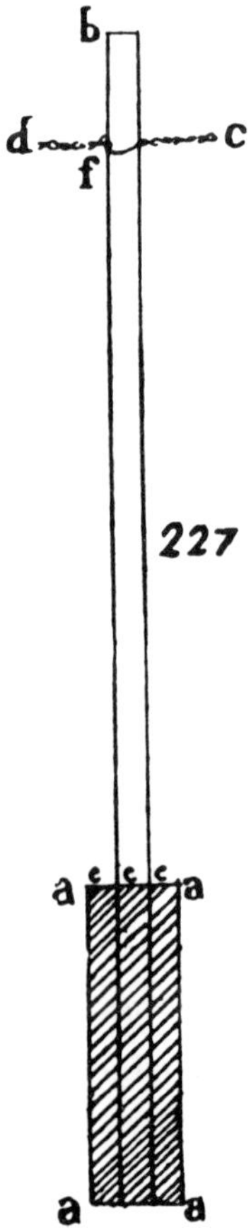

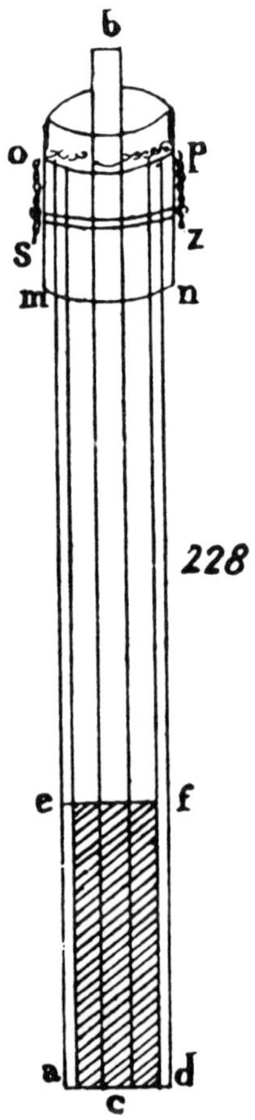

If the tube, fig. 228, were very long, say a foot or more, and it were designed to fill it with metal and to preserve a hole through it, perfectly in the middle, it would be necessary to secure the end, **b**, in some way; it may be managed as follows:

Cut a strip of paper 8 or 9" long and 1 wide, paste it and wind it round the end of the tube in such a manner as to project $\frac{1}{4}$", and form a cup like the top of a hyacinth glass. With a knife or a fine saw, make two slits in the top of this paper cup, as at **o** and **p**, fig. 228, exactly across the middle; take two pieces of thin binding wire 4 or 5" long, and twist them round the top of the rod, as shown at **d f e**, fig. 227; slip them into the slits **o** and **p** of fig. 228; bend them down to **s** and **z**, and tie them in position with a piece of twine, then pour in the melted metal.

The alloy described, 1 regulus of antimony, 4 lead, is type metal; probably some worn-out old type might be obtained of a printer (or wheel weights from a tire seller). Antimony requires a great heat to melt it, but after the lead is mixed with it and cooled, the alloy melts with ease. There is also another alloy which answers equally well, and which in many places may be more easily obtained. Procure some zinc in bars, or some zinc cuttings, or of a zinc chimney-pot; put into a ladle 1 lb. of the zinc, and $\frac{1}{2}$ lb. of lead; melt them; have a piece of tallow as before, or a little powdered sal ammoniac to prevent oxidation; pour into the tubes as before directed. This alloy will be found to be an exceedingly hard metal, almost as white as silver, and will produce splendid drifts.

Regulus of antimony is simply pure antimony--antimony in a metallic state. Regulus is an old alchemical word for the name of the metallic substance that falls to the bottom of the crucible in the melting of ores; as the regulus of antimony, from the reduction of black antimony, antimony sulfide; the regulus of bismuth, arsenic, etc. The old alchemists, whose researches after visionary substances (the philosopher's stone to transmute every metal into gold; elixirs for restoring youth and conferring immortality; panaceas "o'er all the ills of life victorious"; alkahests, or universal solvents, prepared at new and full moon, and thence called menstruums) led to many useful discoveries, esteemed it the purest part of the metal, and gave it the name of regulus, or little king, as being the first-born of the royal metallic race. With type founders the word is still in use, but is employed by them simply as a synonym for antimony; they speak of hardening lead with regulus.

As cases always slightly shrink in drying, and as the drifts previously described are prepared with tubing of the same diameter as the formers, it is obvious that without some preparation it will be impossible to get the drifts into the cases; therefore, either the drifts must be filed smaller, or the formers must be enlarged. Filing the drifts is not only very tedious but it destroys their cylindricity: it is better to enlarge the formers. This may be done by having them turned of soft wood, pine or alder; in large towns drawn tubing may be obtained of the requisite gauge, it should be about 1/32" in diameter more than the drift, ¼ of 1/8. Another way, and which the learner can employ for himself is the following: Take the tube of the same diameter as the drift and paste a piece of writing paper, 10½ diameters wide, to go 3 times round it, tighten it well with the rolling-board and set it by to dry; then with a camelhair brush size it with a coat of thin gum water or isinglass solution; when dry give it a second coat, as soon as that is dry, with another camel-hair brush give it a coat of lac-solution, afterwards a second coat and then a third. This should be done in a warm room, and each coat of the lac-solution ought to have 3 or 4 hours' drying before the next is put on. Roll the cases upon this in the usual way, pasting it or smearing it with a cloth damped with sweet oil, to prevent the brown paper adhering. Wipe it clean before putting it away; at any future time it can have an additional coat of lac-solution if necessary. Cases may also be formed by turning the paper over when pasted, and laying it with the pasted side downwards instead of upwards: place the enlarged former dry upon this and roll up the case, the two dry surfaces coming together, the former will readily deliver and will not require to be previously sized and varnished, but it is not so good a method as the other.

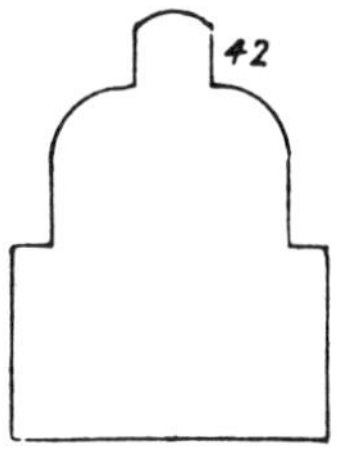

The setting-down piece, fig. 17, may, if preferred, have the hole up it only half an inch, but then it must be driven on a nipple, like fig. 42, of suitable size. It may be either of wood or a tube filled with metal, and the end filed hemispherical.

CONCLUDING REMARKS.

Sublimation is the volatilization of solid substances by heat, and their crystallization by cold again into solids.

The products of sublimation (sublimates) have received the name of flowers from their soft efflorescence, or aggregation of minute spicular crystals into flakes; as flowers of sulfur, the crystallized refrigerated vapour of burning brimstone; flowers of benzoin, benzoic acid; corrosive sublimate, bichloride of mercury; sublimed arsenic, camphor, sal ammoniac; vegetable and lamp black, the condensed fumes of burning oils and resins; soot, the flakes deposited in chimneys from the smoke of burnt wood and coals.

Distillation is the evaporation of liquid substances by heat, and their condensation by cold again into liquids.

The products of distillation (distillates) are usually termed spirits; as spirit of wine; alcohol or brandy; spirit of grain, gin, hollands, or whiskey; spirit of molasses, rum; spirit of naphtha, benzine, etc.

Water, heated or cooled, combines in resemblance the effects of sublimation and distillation; aqueous vapour by congelation crystallizing into snow; and by condensation liquefying into water.

The condensation of steam into water is familiar to everyone. In St. Petersburg, upon the sudden admission of a current of cold air into a crowded assembly room, the vapour in the air was immediately congealed, and fell in the form of snow flakes. To voyagers in the polar regions such occurrence is familiar. The same effect is usual in chambers employed for refrigerating purposes.

Gums are the exudation of trees, vegetable mucilage thickened by exposure to the atmosphere; as gum from cherry and plum trees;

gum arabic, from varieties of the acacia, Turkey, East India, Senegal, or Barbary; Turkey gum arabic is the best.

Resins are the exudation of trees, generally evergreens, essential oils inspissated by oxygenation: as mastic, sandarac, benzoin.

Gums are soluble in water; resins in alcohol and essential oils.

Gums dry and swell up by heat; resins soften and melt.

Gum resins are partly resinous and party mucilaginous; as lac, assafoetida, galbanum. In submitting shellac to the action of alcohol, the whole is never entirely dissolved; as the lac contains, besides the resin, a mucilage which floats about in the liquid, and renders it turbid. Causing the spirit to boil, however, dissolves the great portion.

Many substances which go under the name of gums in commerce, are in reality resins or gum-resins.

Native turpentine, the juice of trees of the fir tribe, of the consistence of honey, yields on distillation spirit of turpentine, called also oil of turpentine, and, by painters, turps; the dry mass left behind in the retort is colophony or rosin. Rosin has been tried in pyrotechny, but is of no use: a solution of it in spirit will bind stars; but it renders them white and smoky. Perhaps it might be used to obtain variety of tint; for which purpose, also, camphor in solution might be employed.

Volatile, ethereal, or essential oils are obtained from plants by distillation with water; as oil of roses, lavender, thyme, peppermint, aniseed, etc. Balsams are mixtures of resins with essential oils.

Fixed oils are obtained from animal fat by heat; and from seeds of plants by pressure and percussion; as train oil, cod-liver oil; palm oil, croton, linseed, cottonseed, etc.

Saw a piece of coke or charcoal in two, and on the flat surface place a few copper filings; direct upon them the flame of a lamp or candle with the blow-pipe; they will simply become red-hot. Lay a few more filings, and on them a little calomel or sal ammoniac: now direct the flame, and a beautiful blue colour will be produced.

Any of the salts of copper may be used to obtain the same effect, the chlorine gas liberated from the calomel (chloride of mercury), or from the sal ammoniac (ammonic chloride) giving a blue colour to all preparations of copper burnt in it.

Weights for quick-match may be made by nearly filling the brass tube with useless rusty old nails, tacks, screws, or odd bits of metal; then pouring in melted lead.

If at any time the basil end of a pin-wheel pipe should be too small to admit the nose of the funnel, it may be enlarged by binding a gum strip round it. If pin-wheel pipes are too dry, they break in winding, from the hardness of the composition; if too damp, from the softness of the paper. As paper cannot be relied upon for being always of uniform thickness, if it be found that a pin-wheel pipe is too thin, cut the strip a little wider. Discretion may be used in all cases.

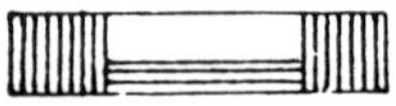

185

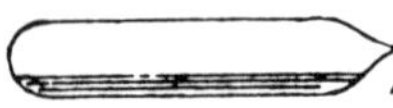

186

To prepare the hollow glass beads, stated to be used by the clown at theatres, procure some imitation grapes, currants or cherries, of an artificial florist, milliner or haberdasher; if grapes or currants, remove the colouring matter by soaking them in water; cherries are lined with wax; warm them by the fire, pull out the stalk, insert a wire, and invert them; the wax will run down the wire; otherwise it will clog. Put in 3 or 4 drops of sulfuric acid, with a pipette, and stop the aperture with melted shellac. Another plan is to enclose the acid in a piece of glass tube, fig. 185. Stop one end with shellac; put in the acid; then stop the other end. If the reader has a Bunsen's burner, the best way of all is to heat the end of the tube and hermetically seal it; put in the acid; heat the tube again, about 1½" from the end; pull it out, and twist it like a spirit-level, fig. 186. Make a mixture of 3 parts of potassium chlorate and 2 of powdered loaf sugar; lay a teaspoonful of it upon a piece of paper; place the charged tube in it, roll up the paper, and twist the ends; lay it upon a slate, or flagstone, and strike it with the flat of a shovel. The apertures must not be stopped with seal-wax, as it is often adulterated with chalk, besides containing colouring substances liable to be attacked by the acid, causing effervesence, and producing cracks. Pieces of broken china may be heated and united with shellac; jet or vulcanite ornaments, earrings, brooches, chains, etc., may be repaired with the same substance.

The best protection against dust is a leather apron, worn with the smooth side inwards; a cotton one is only a very fine sieve.

In charging cases in windy weather, it is advisable to keep the door and windows of the room closed, as the air, at such times, is laden with grit; and, should a particle get blown in, among the composition, the attrition, in driving, might cause an explosion.

It is better to wet plaster of Paris too much than too little; it should always be wetted sufficiently to form a cream, and stirred about in order that the particles may be set free to arrange themselves into the solid form. Gypsum is the native crystalline state of calcium sulfate and contains about 20 per cent of water. When roasted at about 260 degrees, this water of crystallization is expelled, and the mass falls to a white powder, known as plaster of Paris. When this powder is mixed with sufficient water, it sets in a few minutes into a firm mass. The setting is therefore due to the resumption of the combined water expelled by the heat.

If too much brilliant composition should happen to be mixed up, the steel filings can be extracted from it with a magnet; by the same means, iron borings from gerbe compositon. A magnet is also convenient for lifting tacks, screws, etc., from divisions in nail-boxes.

Fig. 169 is a wasp-light, an arsenical suffocating fire for destroying wasps. The proper composition will be found among the powders: drive it into a Roman candle case with the rammer, fig. 4. Bring the leader from the mouth, **b**, backwards along the outside of the case, and tie it in a couple of places, as drawn. Evening is the best time to use it: push the end, **b**, into the nest, light at **a**, and retire.

170

Instead of a five-pointed star, a seven-lance star, fig. 170, may be employed. To form it, have a piece of pine board, $\frac{1}{2}$" thick, $6\frac{1}{2}$" square; draw the diagonals, to find the centre; and with a pair of compasses, stretched to a 3" radius, describe a circle. Carry the radius 6 times round it; and in the points and the centre drive 7 French nails; cut off their heads, and fix on them 7 lances: the middle one, red; the others, 3 green and 3 blue, placed alternately.

For a spindle for small rockets, 3/8" or 4/8", a brass, iron or steel wire, with a few notches filed in it, or made jagged with a cold chisel and hammer, driven into a block, will hold firm without a screw. At Woolwich Arsenal rockets are charged solid. The powder is shaped into pellets, something like large peppermint lozenges, or cylindrical cakes of paste blacking, by hydraulic pressure in a mould. These pellets, discs, or cakes, which are almost as hard as a stone, are put into the case and pressed in tight; the rocket is then fixed upright, and slowly drilled, as I have seen, with a conical borer underneath, working vertically to let the dust fall and clear itself. This mode must not be imitated by an amateur; indeed, without accurate machinery, the desired object could not be effected, and there is constant liability to danger.

The proportions of rockets are:

1/3 cup + 6 choke&core + 1-1/3 solid + 1/3 plug = 8 diameters;

but for 8/8" rockets and larger, 1¼" will be sufficient for the solid.

Rocket spindles are by some called rocket spires.

Chlorate meal, damped with lac-solution or with gum-water, and pressed through a perforated zinc sieve, makes grain-powder. Quickmatch can be made with chlorate-meal, damped with gum-water. The proportion of the ingredients is the same as in gunpowder, but potassium chlorate is substituted for potassium nitrate, to render it miscible with oil or lac-solution. It might be made at the mills.

In the trade mealpowder, potassium nitrate and charcoal go by the names of meal, petre and coal. Common coal, for burning in fires, is never employed in pyrotechny; it would produce only dull red sparks and smoke.

Meal or petre added to a fuse quickens it; sulfur slackens it. 6 meal, 1 sulfur, make a quickmatch that blows through a leader with great violence. 6 meal, 6 sulfur will scarcely burn; pure meal only should be used for match, or grainpowder with hot starch. It has already been stated that potassium nitrate in powder is sometimes adulterated with salt, and that it is impossible to make a rocket with such stuff. Powdered potassium chlorate is sometimes adulterated with nitrate: with such mixture it is equally impossible to produce good colours: nitrate whitens flame and overpowers colour.

Potassium chlorate, charcoal, sulfur, stearine, used separately, with discretion, vivify colours; calomel deepens the colour, but slackens the flame.

Star compositions which inflame vigorously in dry summer weather, will often scarcely burn at all in damp weather; this is especially the case with stars containing strontium nitrate.

In washing sulfur, stir it with a wooden spoon; if a silver one were used, a black sulfide would be formed on the surface, very difficult to remove. If silver coins in the pocket get tarnished while using sulfur, rub them with salt, or chalk, or whiting.

In preparing stars, damp them only just enough to make the composition sticky. If made wet, they dry with difficulty and out of shape; the colour is impaired and very often totally destroyed. In pushing out the star, it ought to fall from the tube, and not stick to the brass rod.

If the brass tube-formers get tarnished, scrape off the lacquer with a knife, sand paper them anew, and give them a fresh coat of lacquer.

Let all wheel-frames and woodwork be coloured black, either with paint, or with a mixture of vegetable or lamp-black and size, or thin glue, or lac-solution to prevent their being seen. A white thread hung upon a bush is visible many yards off; a black one can scarcely be seen a few feet distant. Black is not only invisible, but it throws the brilliancy of sparks, and the vividness of colours into stronger relief.

Let every article be dried, reduced to a fine powder, put into a clean bottle, and carefully corked; also let every bottle be labelled; the labels are best stuck on with paste, not gum: gum labels are apt to drop off in damp weather.

Let all metallic articles liable to rust be wiped with a rag dipped into olive oil before being laid by for future use.

Before putting aside the six inch circular frying-pan, set it over the fire till warm, put into it a lump of tallow, and smear it with a rag; when wanted for use, set it on the fire, put into it a cupful of water and a piece of soda; make the water boil and stir it well round; pour away the water, and dry the pan over the fire.

Let muslin sieves always be dried before being put aside; also again before use. Zinc sieves may simply be wiped dry.

Have a place for everything, and keep everything in its place.

Faraday, the great master in experimental lectures, always devoted many hours to the preparation of his experiments for each lecture. No point, however trifling, bearing upon the success of the experiment, was considered unimportant; he used to try the stoppers of all the bottles he had to use, to see that they had not become fixed, and thus would cause delay by requiring forcible opening. His example cannot be too carefully copied. Before firing a display, all posts, spindles, lines, staples, screws, touch-paper, port-fires, pieces of leadered quickmatch, etc., should be carefully provided. A yard of tape slowmatch, hung to a nail at the top of a post, will supply fire for a long time.

A book should be kept for future guidance, in which should be written the quantity of compositon required to make a certain number of articles of a certain size: by attending to this much waste will be prevented.

Never, upon any account, leave compositions lying about; and let nothing be done by candlelight, except making cases. Quickmatch, especially, ought to be kept locked up, so that nobody can get to it.

Never put squibs, crackers, etc., into the pockets: a stray spark might ignite the whole, and cause most serious mischief.

Never hold large pieces, as gerbes, fixed cases, etc., in the hand; if insufficiently rammed, they will explode. Even in firing squibs or Roman candles it is best to have a glove on the hand. Better to prevent a burn than to cure it.

FORMULATIONS

Roman Candle

Number	1	2	3	4	5	6	7	8	9	10	11	12	13	14
Sulfur	4	2	3	1	1	2	7	3	4	6	8	3	8	-
Charcoal	3	3	3	2	1	3	8	1	1	7	9	3	11	2
Potassium nitrate	8	2	8	4	3	9	21	4	5	18	18	10	32	1
Mealpowder	8	8	3	3	2	4	12	5	4	4	7	-	3	

Rocket

Number	1	2	3	4	5	6	7	8	9	10	11	12	13	14	15	16	17	18
Sulfur	1	1	12	4	8	4	2	4	2	2	1	1	8	1	1	1	2	-
Charcoal	4	2	17	5	11	7	4	8	12	8	2	2	27	2	2	4	4	1
Potassium nitrate	8	5	50	16	32	16	9	16	20	16	4	4	36	4	4	8	8	-
Mealpowder	-	-	-	-	-	-	-	3	1	1	1	2	6	2	1	1	1	4
Steel filings	-	-	-	-	-	-	-	-	-	-	-	-	-	1	1	2	-	

Squib & Serpent

Number	1	2	3
Sulfur	1	-	-
Charcoal	1	1	-
Potassium nitrate	2	-	-
Mealpowder	8	8	4
Steel filings	-	-	1

Chlorate Mealpowder

Number	1	2	3
Potassium chlorate	25	15	60
Charcoal, fine	5	3	9
Sulfur	3	2	8

Wheel & Fixed Cases

Number	1	2	3	4	5	6	7	8	9	10	11	12	13	14	15	16	17
Mealpowder	8	24	8	36	4	18	8	12	42	4	16	10	13	16	20	40	8
Sulfur	1	1	-	1	-	-	1	1	3	-	-	-	1	-	1	4	1
Charcoal	1	4	1	4	1	-	-	-	-	-	-	-	-	-	-	3	2
Potassium nitrate	2	3	-	-	2	2	1	3	8	-	2	-	1	-	-	24	4
Steel filings	-	-	-	-	-	5	3	3	5	1	5	-	-	-	-	6	-
Vegetable black	-	-	-	-	-	-	-	-	-	-	-	1	2	-	-	1	-
Arsenic disulfide	-	-	-	-	-	-	-	-	-	-	-	-	1	-	-	1	-
Black lead oxide	-	-	-	-	-	-	-	-	-	-	-	-	-	3	3	2	-

Starting Fire

Number	1	2	3
Charcoal	1	3	2
Mealpowder	8	16	12
Sulfur	-	4	2
Potassium nitrate	-	6	3

Bursting Powder

Number	1	2	3
Mealpowder	1	1	8
Grain pdr F.	-	1	-
Charcoal	-	-	1

Saxon

Number	1	2	3
Sulfur	1	3	5
Potassium nitrate	1	4	9
Mealpowder	2	9	15

Five-Pointed Star

Number	1	2
Mealpowder	3	-
Sulfur	8	2
Potassium nitrate	12	5
Antimony sulfide	1	1

Spur-Fire for Flower Pots & Star Candles

Number	1	2	3	4	5	6	7
Vegetable black	7	2	3	4	2	4	3
Sulfur	14	5	4	16	6	9	7
Arsenic disulfide	2	1	1	2	1	1	1
Potassium nitrate	32	16	10	32	11	20	15
Mealpowder	-	3	-	17	4	5	4
Charcoal	-	-	-	4	1	1	1

Tourbillion

Number	1	2	3	4	5	6	7	8	9	10	11	12	13	14
Sulfur	1	3	3	7	3	4	4	4	2	4	1	7	4	1
Potassium Nitrate	4	5	16	10	8	32	17	17	4	8	8	20	14	4
Charcoal	2	3	8	4	3	6	4	5	-	-	-	-	3	2
Mealpowder	1	4	8	24	16	32	-	-	4	9	10	23	13	1
Steel filings	-	-	-	8	6	5	-	-	-	-	-	-	2	1
Cast iron borings	-	-	-	-	-	8	6	8	3	4	7	14	5	-

Bengal Light

Number	1	2	3	4	5	6	7
Arsenic disulfide	1	1	2	1	-	-	2
Antimony sulfide	-	-	10	5	1	1	-
Red lead	-	-	1	1	-	-	-
Sulfur	4	2	6	3	4	2	7
Potassium nitrate	14	6	24	14	14	6	24

Inflamant

Number	1	2	3
Charcoal	1	1	1
Mealpowder	8	16	24
FFF grain pdr	4	8	12

Pin Wheel

Number	1	2	3	4	5	6	7	8	9	10	11	12	13
Sulfur	1	3	5	3	7	14	4	4	2	2	3	2	-
Potassium nitrate	1	4	9	5	9	16	8	4	5	2	3	2	-
Mealpowder	2	9	15	8	20	32	26	16	36	26	35	18	3
Antimony sulfide	-	-	-	-	1	3	1	-	-	-	-	-	-
Sawdust, fine	-	-	-	-	-	-	1	-	-	-	-	-	-
Oxalic acid	-	-	-	-	-	-	-	3	-	-	-	-	-
Black lead oxide	-	-	-	-	-	-	-	-	2	-	-	-	-
Arsenic disulfide	-	-	-	-	-	-	-	-	-	2	3	-	-
Vegetable black	-	-	-	-	-	-	-	-	-	1	2	-	-
Lead nitrate	-	-	-	-	-	-	-	-	-	-	-	3	1

Blue Light

Number	1	2
Potassium nitrate	12	5
Sulfur	4	2
Mealpowder	3	-
Antimony sulfide	1	1

Wasp Light

Number	1
Potassium nitrate	14
Sulfur	4
Mealpowder	3
Arsenic disulfide	1

Port-Fire

Numbers	1	2	3	4
Mealpowder	3	1	1	5
Sulfur	4	2	2	8
Potassium nitrate	12	4	6	25

Shell-Fuse

Number	1
Mealpowder	4
Potassium nitrate	2
Sulfur	1

Gerbe

Number	1	2	3	4	5	6	7	8	9	10	11	12
Sulfur	2	3	-	1	2	3	2	2	2	-	-	
Potassium nitrate	2	2	-	-	-	10	6	4	9	4	-	-
Mealpowder	16	36	4	8	16	-	9	16	2	2	8	3
Steel filings	1	6	-	-	-	-	-	-	-	-	-	-
Cast iron borings	5	8	1	3	7	5	8	5	3	-	-	
Charcoal	-	-	-	-	1	2	2	1	2	2	-	-
Coke grains	-	-	-	-	-	-	-	-	-	-	1	-
Porcelain grains	-	-	-	-	-	-	-	-	-	-	-	1

Tailed, Streamer, or Comet Stars for Rockets, Shells & Roman Candles

Number	1	2	3	4	5	6	7	8	9	10	11	12	13	14
Vegetable black	1	3	3	6	-	-	-	-	3	1	-	-	-	-
Charcoal	2	8	2	3	1	3	6	12	2	4	1	1	3	-
Sulfur	5	24	2	-	2	4	1	2	2	12	3	3	8	-
Potassium nitrate	5	24	9	-	-	-	10	20	9	12	4	4	4	-
Mealpowder	8	30	6	16	-	20	5	7	6	8	3	3	12	2
Sodium oxalate	-	-	-	-	4	12	-	-	-	-	-	-	-	-
Antimony sulfide	-	-	-	-	-	-	-	-	4	-	-	-	-	-
Potassium chlorate	-	-	-	-	8	16	-	-	-	-	-	-	-	-
Asphaltum	-	-	-	-	-	-	-	-	-	-	1	-	-	-
Burgundy pitch	-	-	-	-	-	-	-	-	-	-	-	1	-	-
Coke grains, fine	-	-	-	-	-	-	-	-	-	-	-	-	4	1

Oiled Tailed Stars for Rockets & Shells

Number	1	2	3
Charcoal	9	6	2
Sulfur	9	5	2
Potassium nitrate	32	18	9
Mealpowder	24	12	5
Antimony sulfide	16	9	4

To 1 ounce add twenty-four drops of boiled linseed oil; rub them thoroughly together in a mortar; then spread out the mixture, for a few days, to dry. When dry, mix with starch, dextrine solution, or gum-water, and chop into 3/8" or ½" cubical blocks. They will keep for years, and improve by age. In order that a star may tail, it must rapidly burn through and leave a cinder, or scoria; from this, as it falls, minute portions become detached, and trail behind.

Steel Stars for Rockets & Shells

Number	1	2	3	4	5
Lead Nitrate	8	24	28	-	-
Potassium chlorate	3	5	6	5	4
Charcoal	2	6	6	1	-
Steel filings	2	6	6	3	4
Potassium nitrate	-	4	3	-	-
Shellac	-	-	1	-	-
Sulfur	-	-	-	1	1

Rub up the mixture thoroughly in a mortar, with just enough boiled oil to make it cohere, and pump it into Roman candle stars; the oil will preserve the steel from rusting. For Roman Candles, or Italian Streamers, they will be ready at any time; for Rockets and Shells they may be matched and enveloped, like fig. 32, a day or two previously. They form beautiful stars. Or they may be charged in cases, and primed with chlorate mealpowder. Or they may be damped with lac-solution.

Pearl Streamer

Number	1	2	3	4	5
Potassium nitrate	12	26	2	-	-
Charcoal	5	11	1	-	-
Zinc filings	14	28	4	10	15
Mealpowder	-	-	1	8	12
Vegetable black	-	-	-	1	1

Instead of filings, zinc may be obtained in a fine powder, by pouring it, melted, into a hot iron mortar, and hammering it with the pestle directly it begins to solidify. Sift it through a fine sieve. Protect the hands with cloth gloves while using the pestle. Damp the composition with gum-water for Roman candle stars. Broken bits of the stars may be put into coloured Gerbes.

Gold Rain

Number	1	2	3	4	5
Sulfur	1	1	1	-	-
Potassium nitrate	2	2	2	-	-
Charcoal	6	1	5	1	3
Mealpowder	16	6	18	4	8

Silver Rain

Number	1	2	3	4
Steel filings	1	2	2	3
Mealpowder	4	7	8	-
Potassium nitrate	-	1	-	2
Sulfur	-	-	1	-
Charcoal	-	-	-	3
Lead nitrate	-	-	-	10

White or Bright Stars & Lances

Number	1	2	3	4	5	6	7	8	9	10	11	12	13	14	15
Mealpowder	3	3	2	1	1	1	1	1	3	-	-	-	-	-	-
Sulfur	4	6	4	2	3	2	2	1	11	5	7	-	8	2	1
Antimony sulfide	1	-	1	-	3	-	2	1	10	5	6	3	1	1	2
Arsenic disulfide	-	-	-	1	-	1	-	-	-	3	5	-	-	-	-
Red lead	-	-	-	-	-	-	-	-	-	2	-	-	-	-	-
Lead nitrate	-	-	-	-	-	-	-	-	-	-	-	3	16	3	12
Potassium chlorate	-	-	-	-	-	-	-	-	-	-	-	28	16	4	3
Shellac	-	-	-	-	-	-	-	-	-	-	-	5	-	-	1
Potassium nitrate	8	14	14	5	10	6	9	4	48	24	34	18	-	-	-

Sugar Blues for Stars & Lances

Number	1	2	3	4	5	6	7	8	9	10
Potassium chlorate	8	36	40	40	36	9	44	40	6	9
Mercurous chloride	4	18	24	24	12	3	12	-	5	5
Sugar	3	12	9	12	4	1	12	9	4	3
Copper sulfate	5	22	22	12	4	3	12	22	-	-
Stearine	-	1	2	1	3	1	1	2	-	-
Copper oxychloride	-	-	-	3	4	2	5	-	-	-
Ammonium chloride	-	-	-	-	-	-	-	6	-	-
Copper filings	-	-	-	-	-	-	-	-	1	-
Black copper oxide	-	-	-	-	-	-	-	-	-	2

See explanation on next page.

No substance combines better with salts of copper than sugar. Sugar, put into the bowl of a tobacco pipe and placed in the fire, burns fiercely, and is converted into caramel. This poured on to a plate, slightly smeared with butter to prevent it sticking, hardens on cooling; and is used for colouring brandy, vinegar, gravy, porter, coffee, etc. Stearine must be scraped very fine from a stearine candle. Sugar blues are to be damped with pure water only, as the sugar itself, when wetted, is sufficiently cohesive. Use an exceedingly small quantity of water, and rub it up thoroughly in the mortar; the longer it is rubbed, the better it combines.

Crimson & Scarlet Stars & Lances

Number	1	2	3	4	5	6	7	8	9	10	11	12	13	14	15	16	17	18	19	20	21	22	23	24
Potassium chlorate	16	8	16	16	24	16	24	16	8	16	25	4	32	6	16	28	32	26	96	8	24	8	2	13
Strontium nitrate	16	16	32	32	-	16	20	24	5	6	30	7	48	5	-	-	42	10	72	12	18	12	8	32
Sulfur	5	6	9	12	6	5	10	-	1	-	10	1	6	-	-	-	13	5	24	2	-	2	2	8
Charcoal, fine	1	1	1	1	1	-	-	-	-	-	-	-	-	-	-	-	-	-	-	-	-	-	1	2
Shellac	-	1	4	2	2	-	-	7	1	4	3	1	12	1	-	-	4	3	21	4	5	3	-	-
Mercurous chloride	-	-	7	-	6	1	-	14	2	2	9	2	12	5	-	12	12	10	42	-	-	4	-	-
Copper sulfate	-	-	3	1	-	-	-	1	1	1	3	1	-	1	-	-	4	1	4	1	5	-	-	-
Arsenic disulfide	-	-	-	-	-	1	5	1	-	-	-	-	-	-	-	-	-	-	-	-	1	-	-	-
Vegetable black	-	-	-	-	-	-	-	-	-	-	1	-	1	-	-	-	1	-	-	-		1	-	-
Sugar	-	-	-	-	-	-	-	-	-	-	-	-	-	-	7	12	-	-	-	-	-	-	-	-
Strontium carbonate	-	-	-	-	10	-	-	-	-	-	-	-	-	-	11	5	-	-	-	-	-	-	-	-

Blue Stars & Lances Without Sugar

Number	1	2	3	4	5	6	7	8	9	10	11	12	13	14	15	16	17	18	19	20	21	22	23
Potassium chlorate	5	40	18	40	6	8	48	40	24	16	30	24	8	22	40	6	16	24	2	5	40	40	36
Mercurous chloride	4	20	8	28	2	2	12	12	6	8	10	8	1	8	20	3	1	8	-	-	-	-	12
Copper sulfate	4	20	10	28	-	2	-	-	-	-	-	-	3	6	25	5	-	-	-	1	20	22	-
Shellac	1	5	-	-	-	-	2	2	1	-	-	1	-	-	5	-	-	1	-	-	5	-	6
Copper oxychloride	-	2	-	-	-	-	8	9	6	5	10	4	1	2	-	-	2	-	1	1	-	-	9
Dextrine	-	-	5	1-	-	-	-	-	-	-	-	-	-	-	2	2	-	-	-	-	-	-	-
Sulfur	-	-	-	-	3	4	4	-	1	2	3	2	3	5	-	-	4	2	1	2	5	5	9
Stearine	-	-	-	3	-	-	1	1	2	2	3	1	-	-	-	-	-	-	-	-	-	3	1
Black copper oxide	-	-	-	-	1	1	1	-	-	-	-	-	-	-	-	-	-	-	-	-	-	-	-
Copper filings	-	-	-	-	-	-	-	-	-	-	-	-	-	-	-	-	-	5	-	-	-	-	-
Ammonium chloride	-	-	-	-	-	-	-	-	-	-	-	-	-	-	-	-	-	-	-	-	6	6	-

It is impossible to powder shellac sufficiently fine by hand; and, twenty years ago, powdered shellac could not be procured. About that time the drug-grinders, finding a demand for it, submitted it to the action of the stamping mills (mechanical pestle and mortar), and now it can be obtained at most shops.

Chertier mixed flake shellac with salt; melted the two together; powdered the mixture; and washed out the salt. Such process is needless now. It is useless, unless as fine as wheaten flour.

Green or Emerald Stars & Lances

Number	1	2	3	4	5	6	7	8	9	10	11	12	13	14	15	16	17	18	19	20	21	22
Potassium chlorate	16	8	132	32	144	8	3	1	-	16	6	12	16	3	48	22	24	108	24	16	2	13
Barium nitrate	16	8	108	54	160	21	2	-	-	8	7	5	8	3	42	22	32	108	32	48	10	32
Barium chlorate	-	-	-	-	-	3	2	2	4	8	3	4	4	-	-	-	-	-	-	-	-	-
Sulfur	5	4	6	6	4	7	1	1	1	5	-	-	1	-	-	-	10	18	8	12	2	8
Charcoal, fine	1	1	-	-	-	1	-	-	-	1	-	1	-	-	-	-	-	-	-	1	1	2
Antimony sulfide	-	1	-	-	-	-	-	-	-	-	-	-	-	-	-	-	-	-	2	-	-	-
Mercurous chloride	-	-	48	27	100	-	-	-	-	-	5	2	2	2	-	10	-	48	-	8	-	-
Shellac	-	-	24	12	12	2	-	-	-	-	-	3	4	-	7	1	-	24	-	5	-	-
Vegetable black	-	-	1	1	1	-	-	-	-	-	-	-	-	-	-	-	-	1	-	-	-	-
Sugar	-	-	-	-	-	-	-	-	-	-	5	-	-	2	7	14	-	-	-	-	-	-
Ammonium chloride	-	-	-	-	-	-	-	-	-	-	-	-	-	-	5	-	-	-	-	-	-	-
Arsenic disulfide	-	-	-	-	-	-	-	-	-	-	-	-	-	-	-	-	3	-	3	-	-	-
Copper sulfate	-	-	-	-	-	-	-	-	-	-	-	-	-	-	-	-	-	-	-	2	-	-

If powdered barium nitrate and shellac crushed by being hammered in a bag, are mixed together and melted in a pipkin over the fire, the mixture, when cold, may be reduced to a powder in an iron mortar, with patience. Take Number 6. Weigh out 21 parts barium nitrate, and 2 parts of coarsely powdered lac; melt them together; when cold, powder them; and add the other substances in proper proportion. Shellac may be melted with strontium nitrate, in the same way.

DEEP AND PALE YELLOW STARS & LANCES

Number	1	2	3	4	5	6	7	8	9	10	11	12	13	14	15	16	17	18	19
Potassium chlorate	8	4	12	16	12	16	4	4	16	3	8	16	16	4	8	6	20	5	6
Sodium oxalate	3	2	8	4	4	4	3	1	4	4	4	5	-	1	-	4	15	-	3
Sodium bicarbonate	-	-	-	-	-	-	-	-	-	-	-	-	3	-	3	-	-	1	-
Strontium nitrate	-	-	-	-	-	-	10	-	-	-	16	-	4	-	-	-	-	-	-
Strontium carbonate	-	-	-	3	2	3	-	-	-	-	-	-	-	-	-	-	-	-	-
Barium nitrate	-	-	-	-	-	-	-	-	4	10	-	-	3	-	-	7	30	-	-
Sulfur	-	-	-	4	4	4	-	1	2	-	6	5	-	-	-	3	8	1	2
Shellac	2	1	3	-	1	1	5	-	2	3	1	-	4	1	-	1	4	-	1
Stearine	-	-	-	-	-	1	-	-	1	-	-	-	-	-	-	-	-	-	-
Charcoal, fine	.	-	-	-	-	-	-	-	-	-	1	1	-	-	-	-	1	-	-
Arsenic disulfide	-	-	-	1	1	1	-	-	-	-	-	-	-	-	1	-	-	1	-
Sugar	-	-	-	-	-	-	-	-	-	-	-	-	-	1	3	-	-	-	-

MAUVE AND LILAC STARS & LANCES

Number	1	2	3	4	5	6	7	8	9	10
Potassium chlorate	28	17	60	40	25	24	24	25	12	6
Mercurous chloride	12	-	-	-	10	12	12	-	-	2
Shellac	4	-	-	-	5	5	5	-	-	1
Strontium nitrate	4	4	25	14	-	4	-	16	16	1
Copper sulfate	2	7	20	-	5	2	2	-	-	1
Stearine	1	-	-	-	1	-	1	-	-	-
Sulfur	-	7	35	16	-	-	-	12	2	-
Lead chloride	-	1	-	2	-	-	-	-	-	-
Lead nitrate	-	-	-	-	10	-	12	1	-	-
Copper Oxychloride	-	-	8	12	-	-	-	6	-	-
Ammonium chloride	-	-	1	-	-	-	-	-	-	-
Vegetable black	-	-	-	1	-	-	-	1	1	-
Potassium nitrate	-	-	-	2	-	-	-	2	1	-
Strontium carbonate	-	-	-	-	5	-	4	-	-	-
Arsenic disulfide	-	-	-	-	-	-	1	-	1	-

PURPLE AND VIOLET STARS & LANCES

Number	1	2	3	4	5	6	7	8	9	10	11	12	13	14	15
Potassium chlorate	42	28	48	16	6	16	3	6	26	30	96	24	20	32	37
Strontium nitrate	42	14	48	-	4	-	-	1	-	-	24	-	-	-	-
Sulfur	13	-	28	2	1	6	1	3	-	3	-	2	6	12	9
Mercurous chloride	12	14	28	7	2	6	2	20	8	48	8	8	12	13	
Copper sulfate	4	1	40	8	1	-	3	-	3	12	1	-	-	-	-
Shellac	4	5	1	-	1	-	-	-	-	4	-	1	1	2	8
Vegetable black	1	-	-	-	-	-	-	-	-	-	-	-	-	-	-
Black oxide of copper	-	-	-	-	-	4	4	1	-	-	-	-	-	1	1
Strontium carbonate	-	-	-	-	-	-	-	-	4	12	-	-	-	-	-
Sugar	-	-	-	-	-	-	-	-	14	-	42	-	-	-	-
Copper oxychloride	-	-	-	-	-	-	-	-	-	-	-	4	5	8	9
Stearine	-	-	-	-	-	-	-	-	-	-	-	2	2	-	1

MAGNESIUM COLORS FOR STARS & ASTERIODS

Colors	Bright Red	Dark Red	Green	Blue	Yellow	White
Strontium nitrate	8	6	-	-	-	-
Barium chlorate	-	-	12	-	-	-
Copper oxychloride	-	-	-	2	-	-
Sodium oxalate	-	-	-	-	2	-
Antimony sulfide	-	-	-	-	-	1
Potassium chlorate	2	4	-	5	4	-
Potassium nitrate	-	-	-	-	-	12
Sulfur	2	2	1	2	-	4
Charcoal	1	-	-	-	-	-
Shellac	-	2	3	1	1	-
Mercurous chloride	-	4	-	2	-	-
Magnesium powder	2	3	2	2	1	2

A few magnesium-filings may be added to any colour.

SLOW FIRES TO BE HEAPED UPON A TILE IN SHAPE OF A CONE AND LIT AT TOP

Colors	Red	Red	Red	Green	Green	Green	Purple	Purple	Yellow	Yellow	Crimson	Crimson
Strontium nitrate	16	24	108	-	-	-	108	72	20	-	40	36
Barium nitrate	-	-	-	16	16	16	-	-	-	10	-	-
Sodium oxalate	-	-	-	-	-	-	-	-	3	5	-	-
Copper Sulfate	-	3	30	-	-	-	24	3	-	-	-	-
Barium chlorate	-	-	-	-	-	12	-	-	-	-	-	-
Potassium chlorate	1	3	12	1	1	-	9	4	2	2	5	4
Charcoal, fine	1	-	-	1	-	-	-	-	-	-	2	-
Mercurous chloride	-	6	24	-	5	9	24	18	-	-	-	-
Sulfur	4	8	39	4	2	7	39	24	4	1	13	12
Shellac	-	1	2	-	2	1	2	3	2	6	-	1
Vegetable black	-	-	1			-	1	2	-	-	-	1
Antimony sulfide	-	-	-	-	-	-	-	-	-	-	4	4

In order to get the powder into a conical heap, press it into a wine-glass, or into fig. 65, lay a tile upon the top, and invert.

1

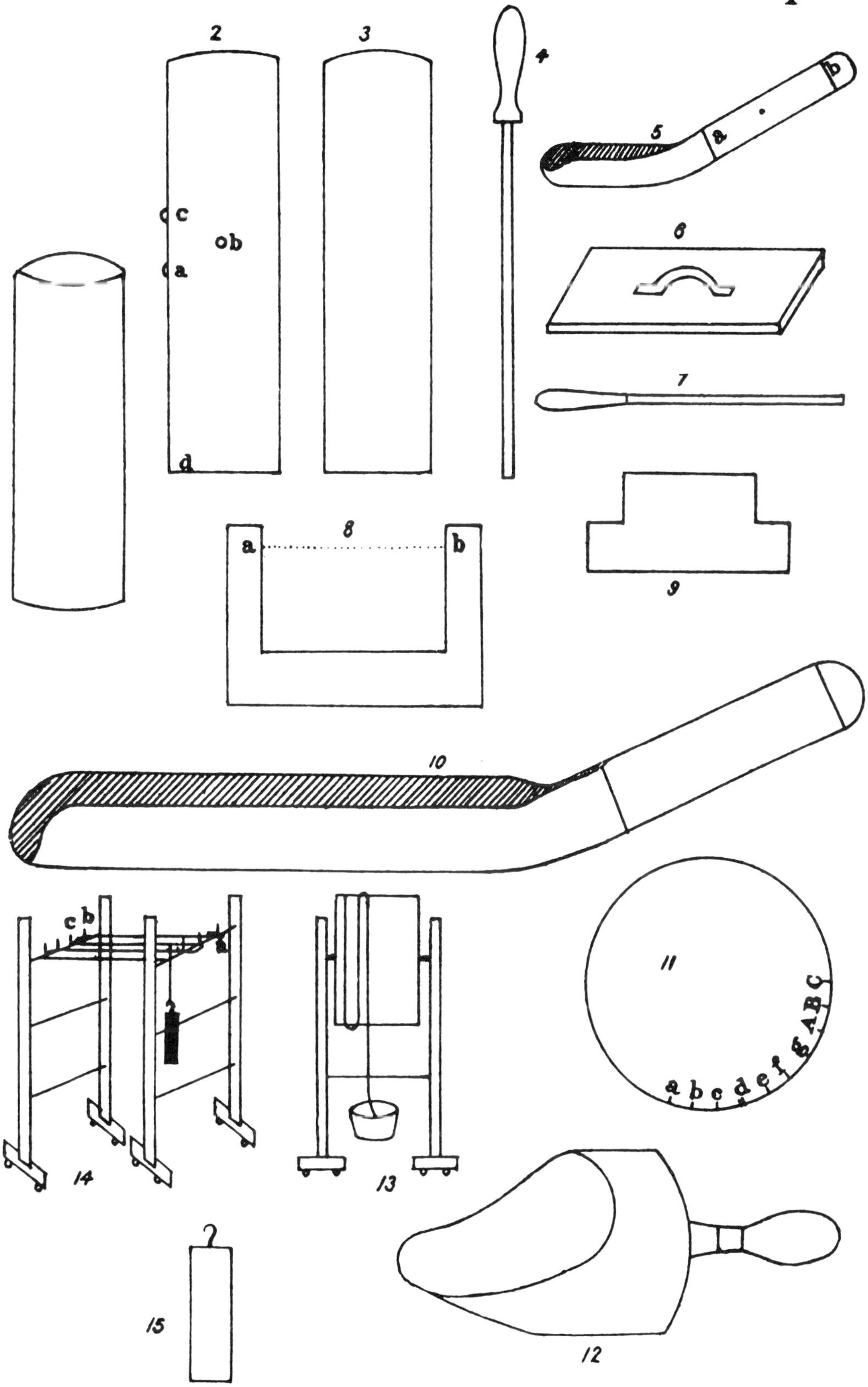

2

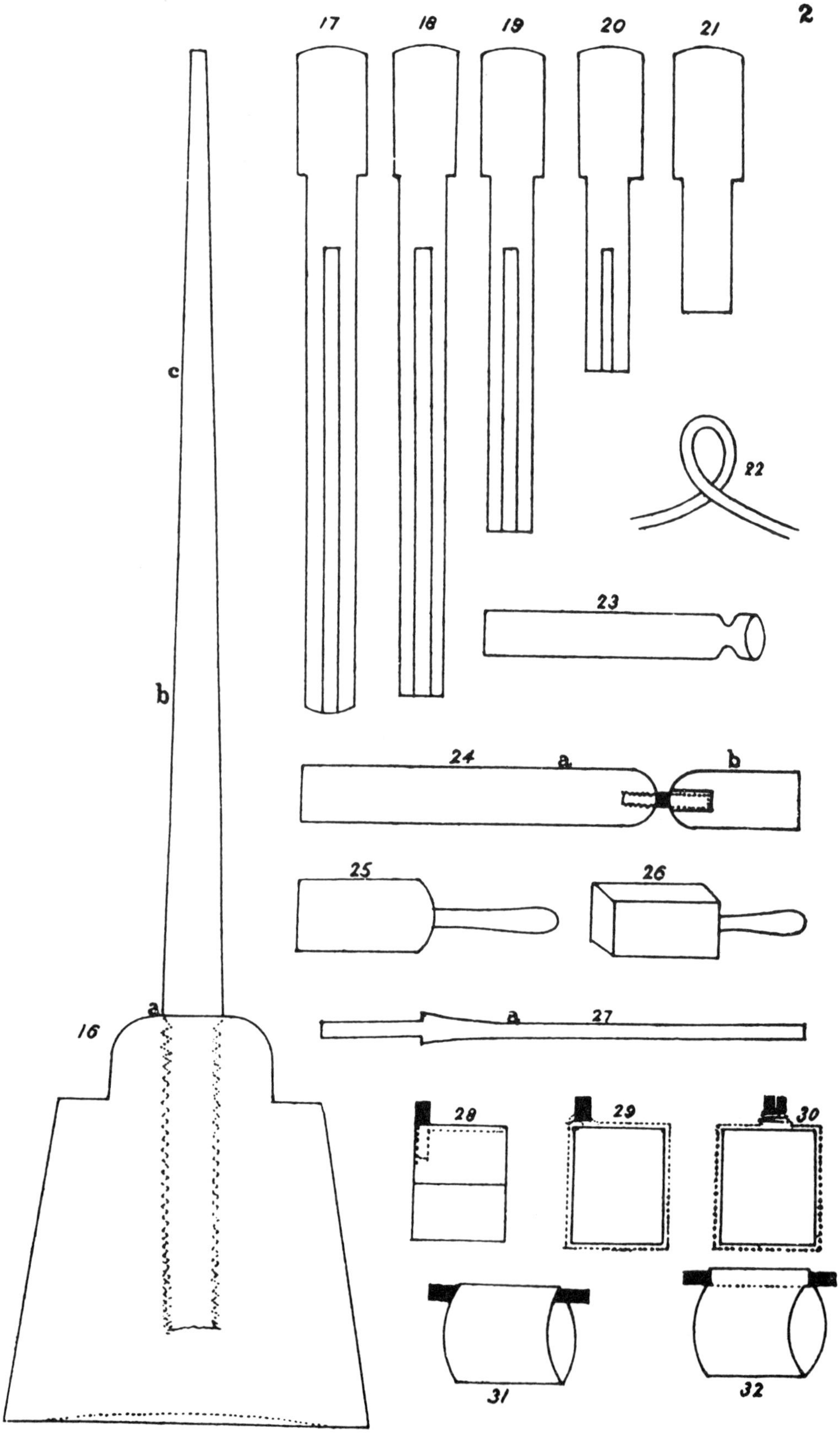

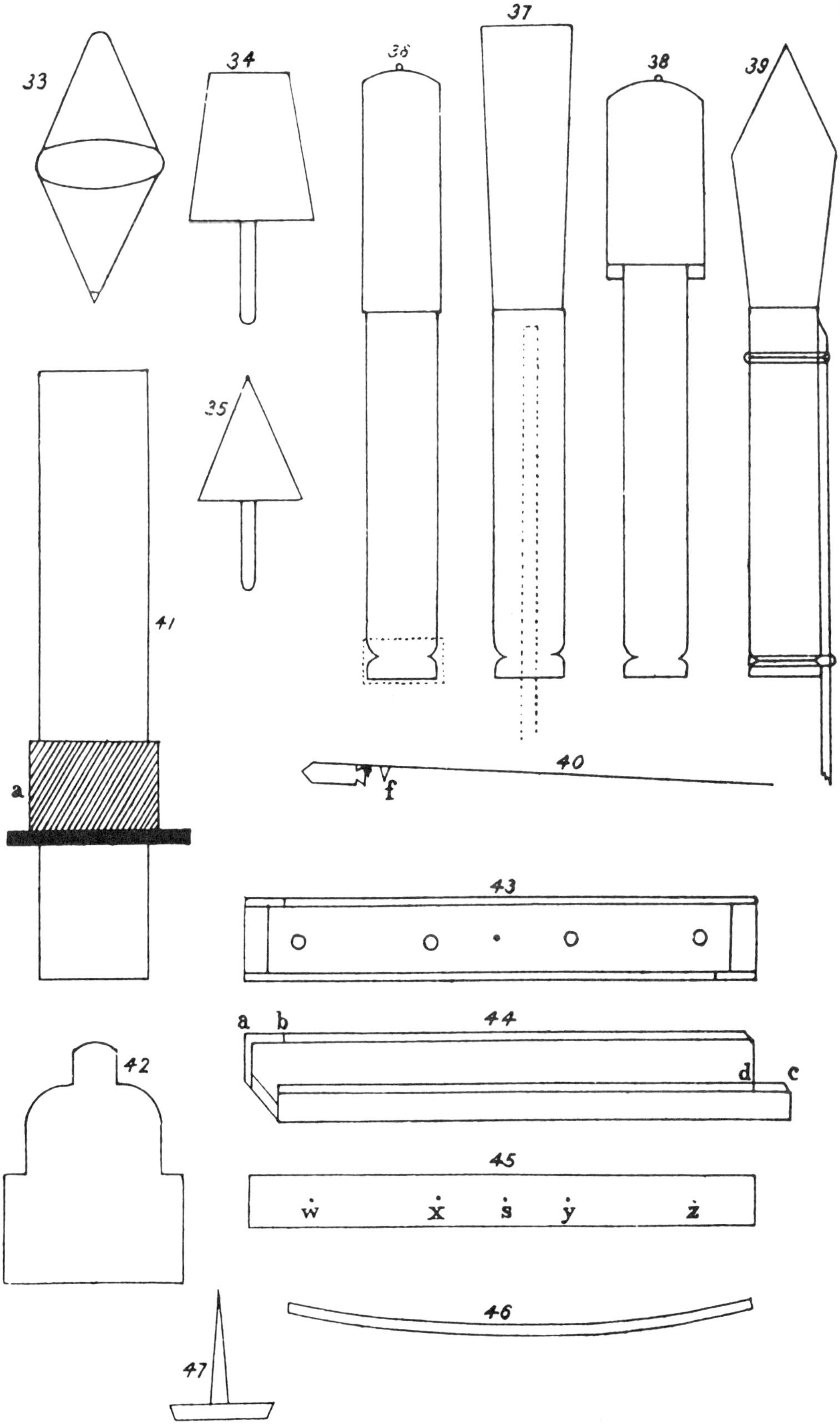
33
34
36
37
38
39
35
41
a
40
f
43
44
a
b
d
c
42
45
w
x
s
y
z
46
47

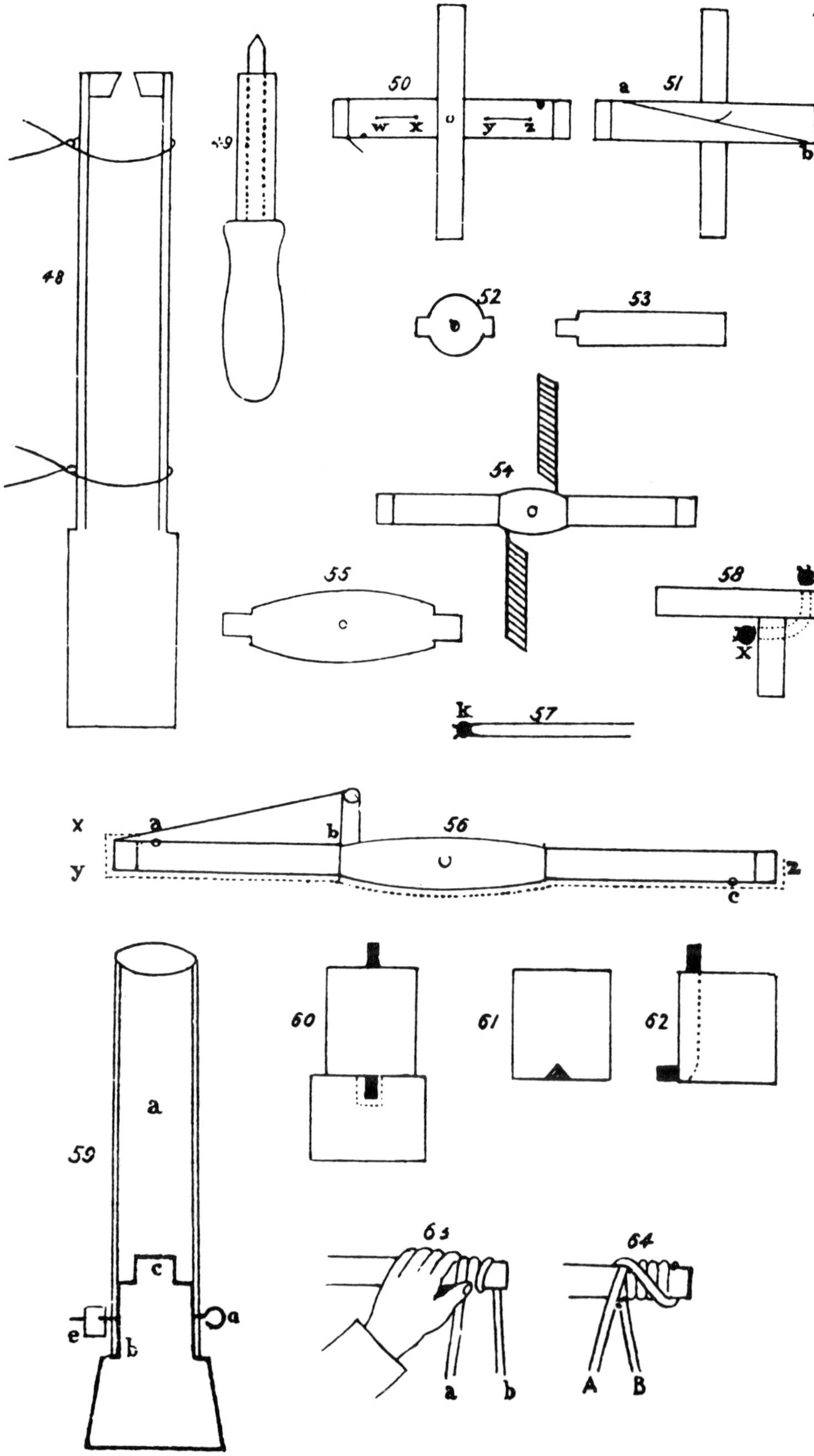
48
49
50
w x y z
51
a b
52
53
54
55
58
z x
57
k
56
x a b y z c
59
a c e b a
60
61
62
63
a b
64
A B

65

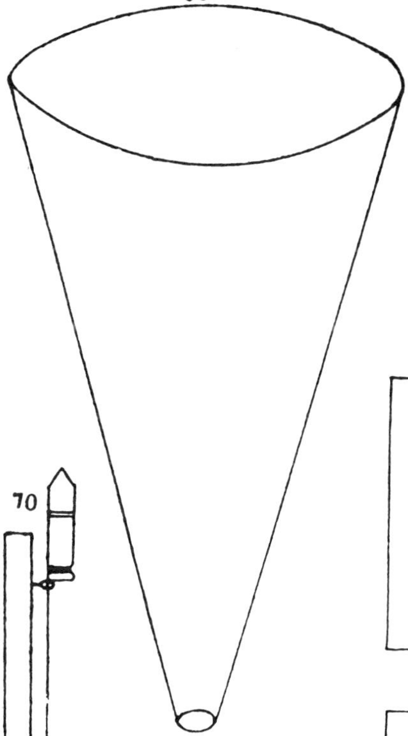

66

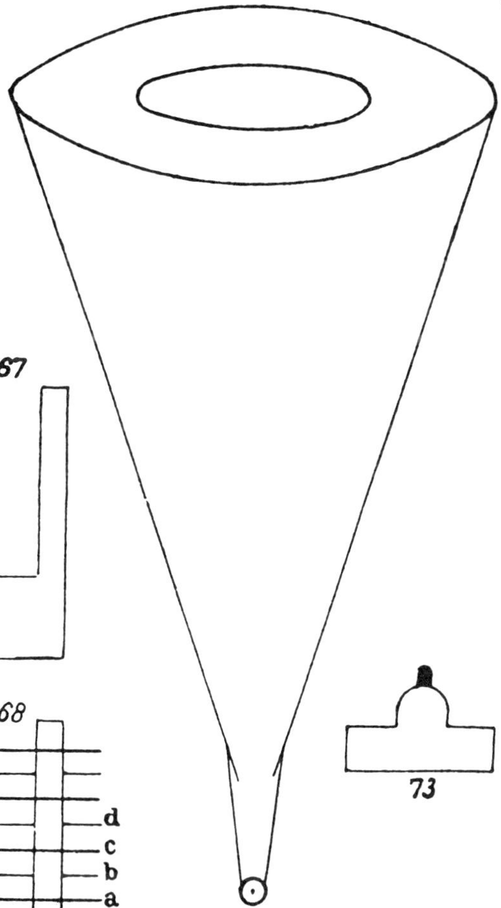

67

70

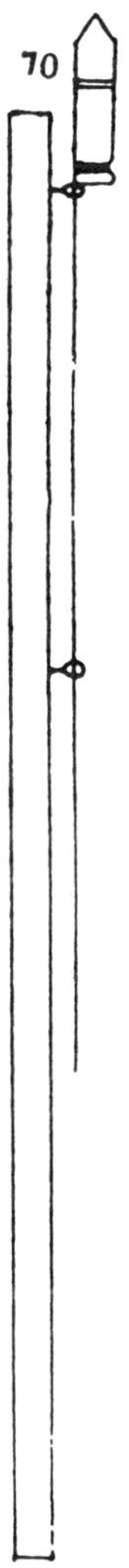

73

68

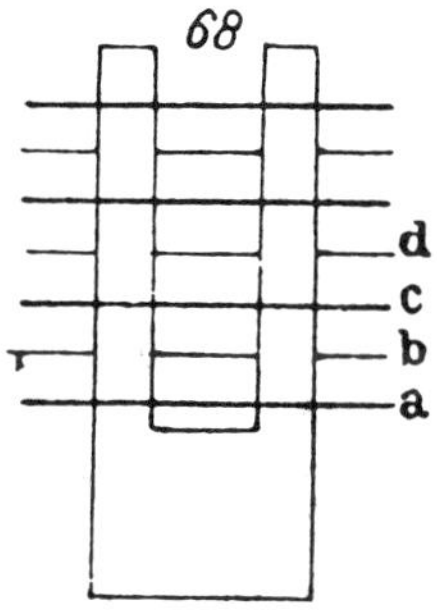

69

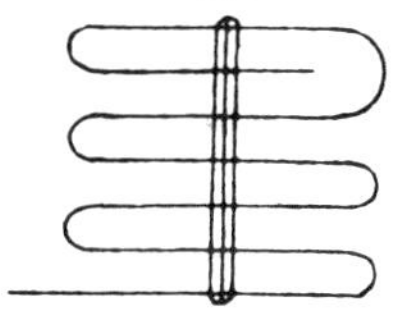

71

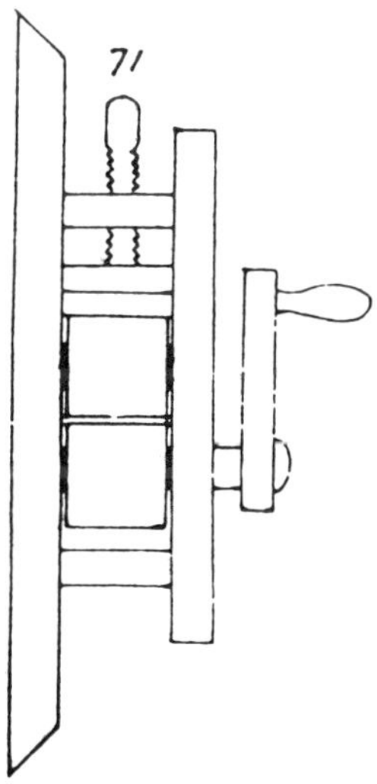

72

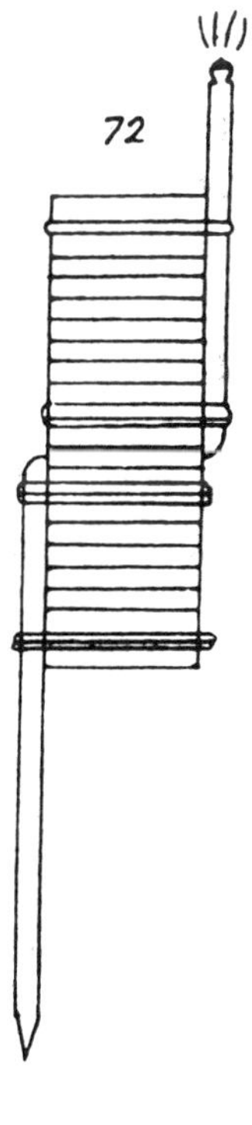

74

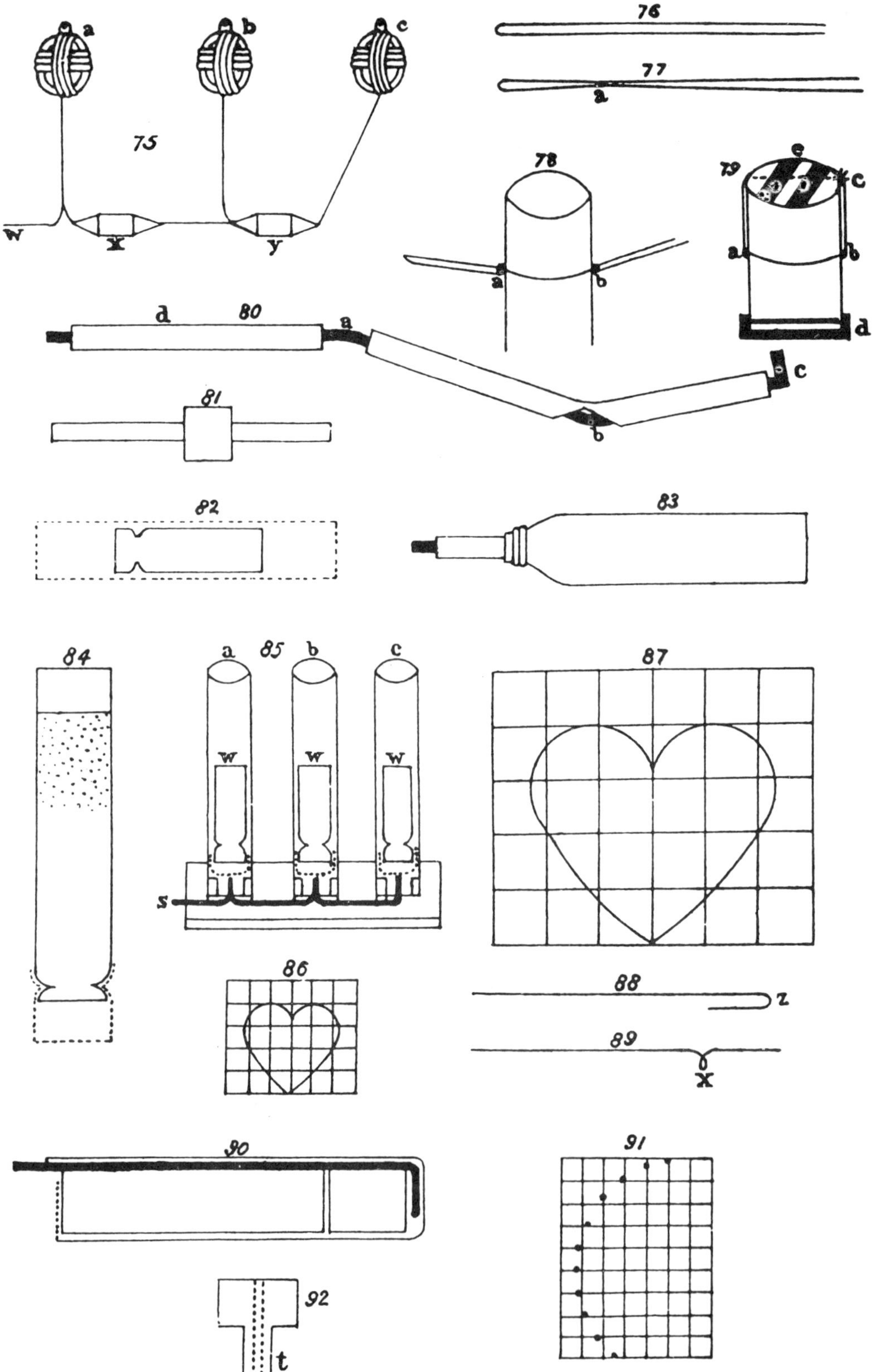
75
a
b
c
w
x
y
76
77
a
78
a
b
79
e
c
a
b
d
80
d
a
b
c
81
82
83
84
85
a
b
c
w
w
w
s
86
87
88
z
89
x
90
91
92
t

93

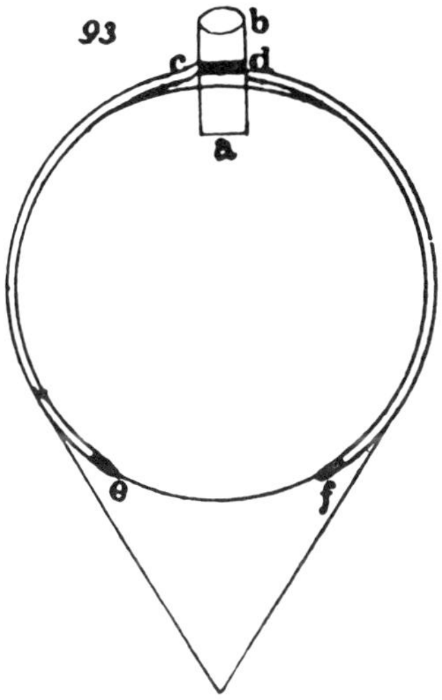

94

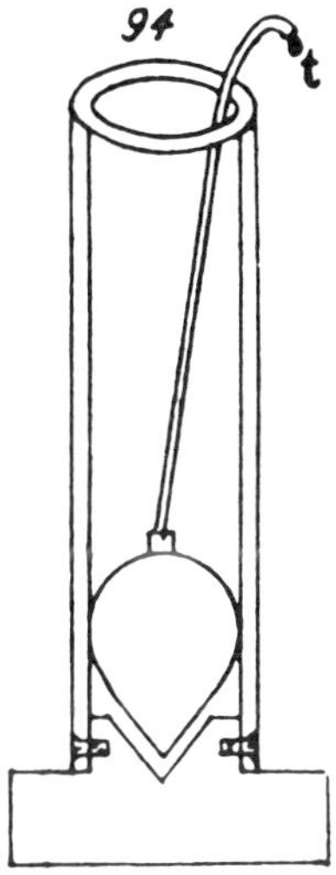

95

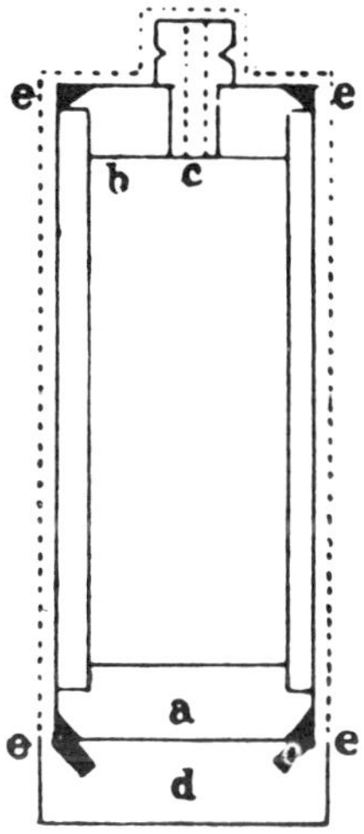

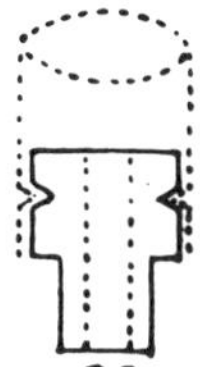

96

99

97

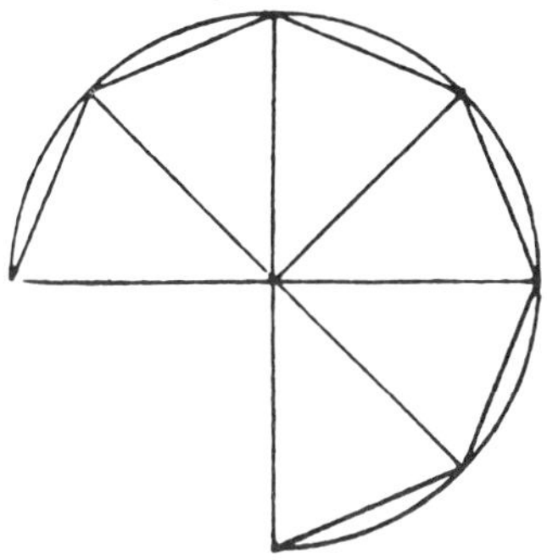

98

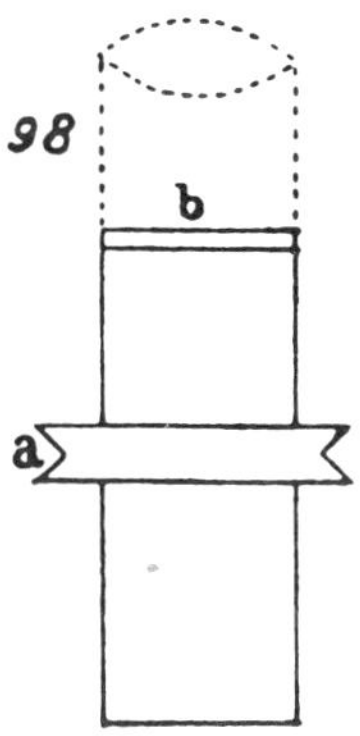

100

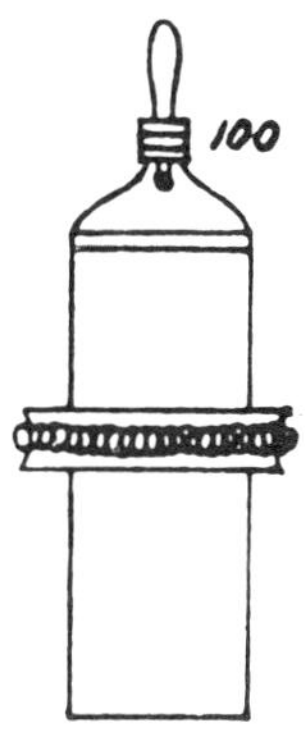

101

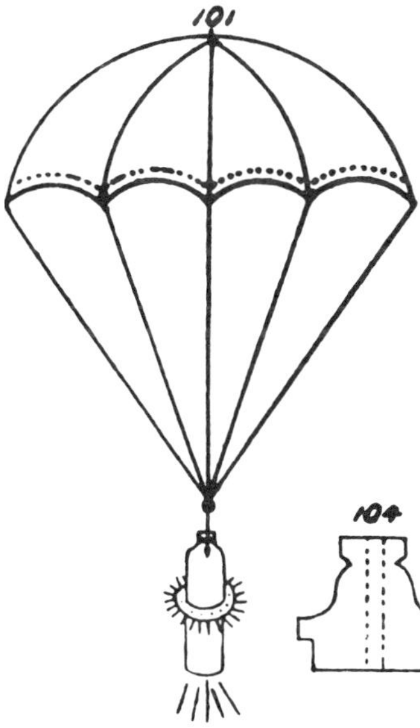

102

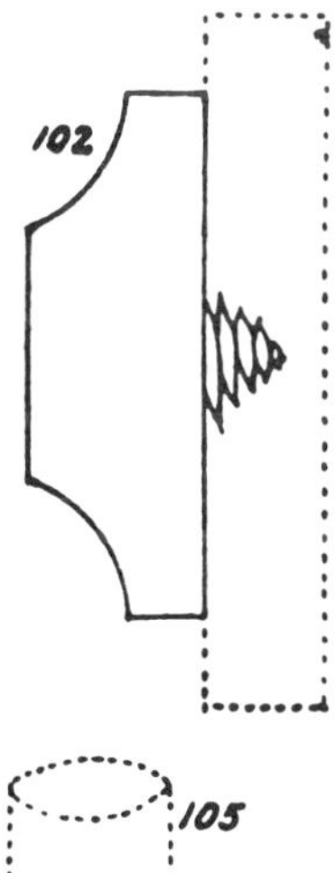

103

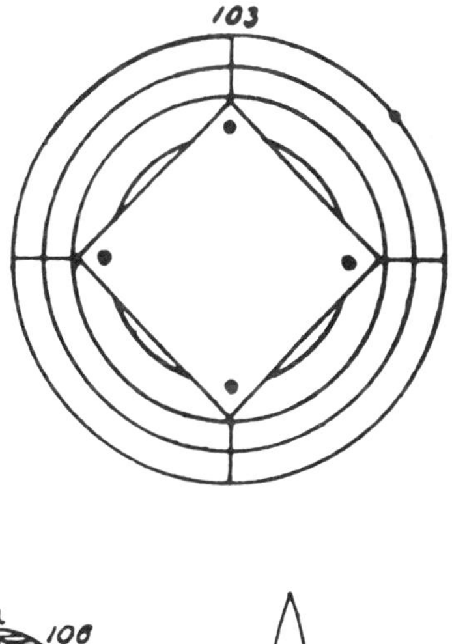

104

105

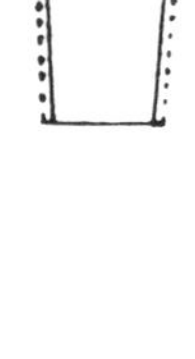

106

107

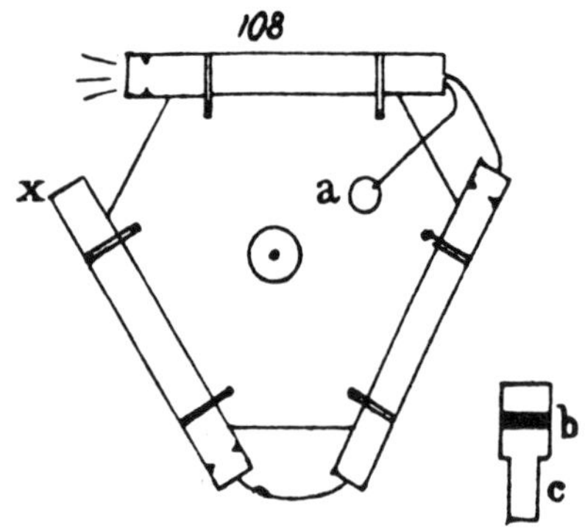
108
x
a
b
c

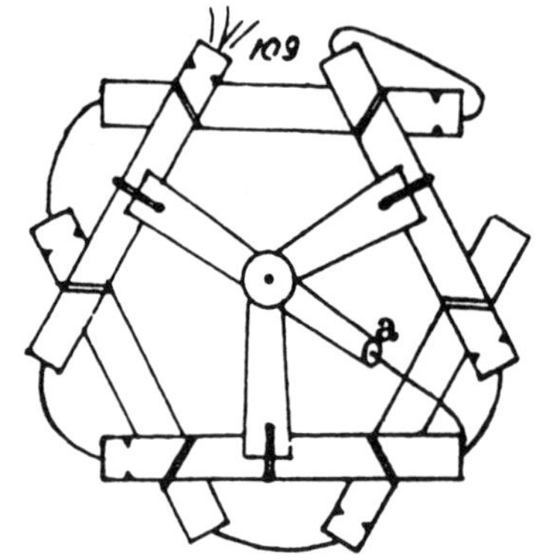
109
a

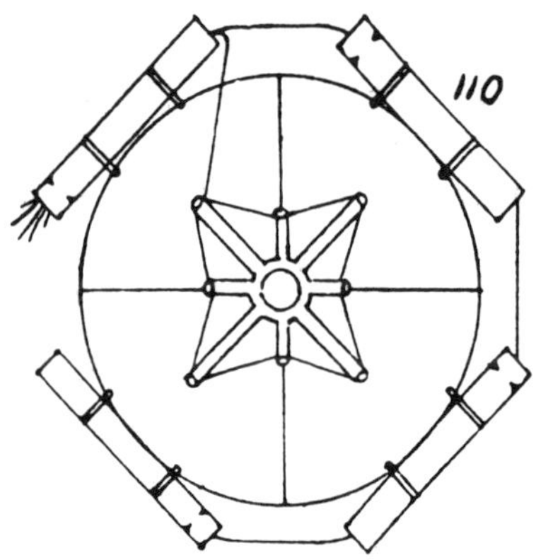
110

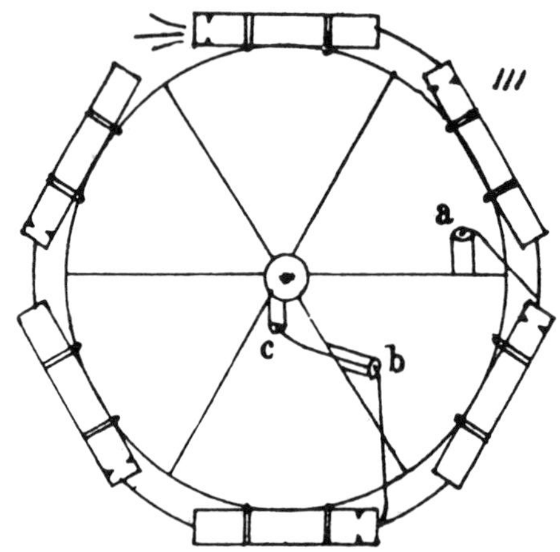
111
a
c
b

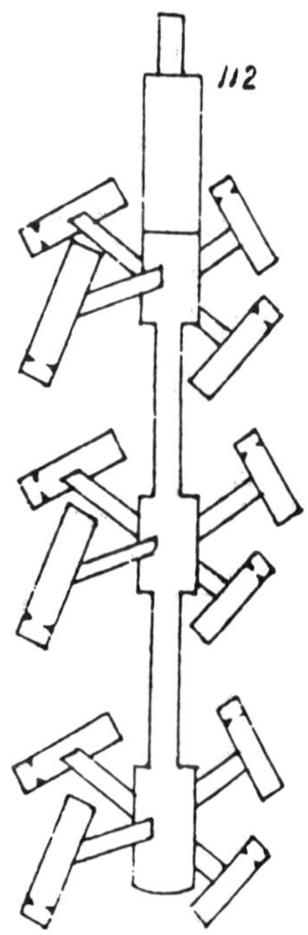
112

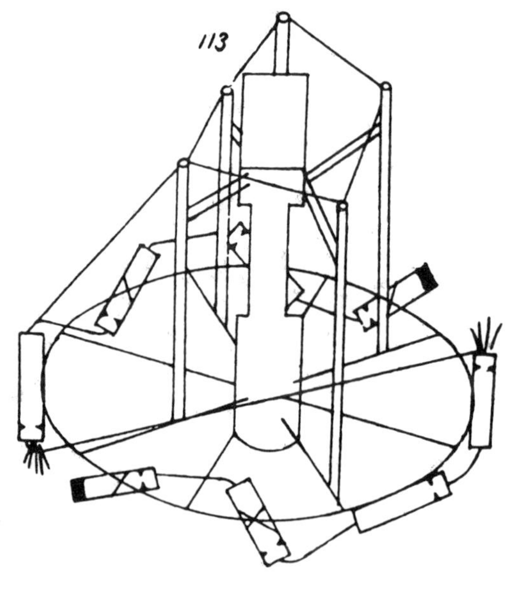
113

114

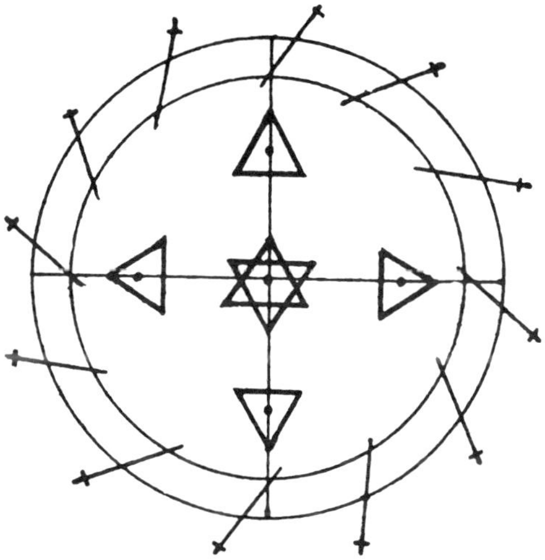

115

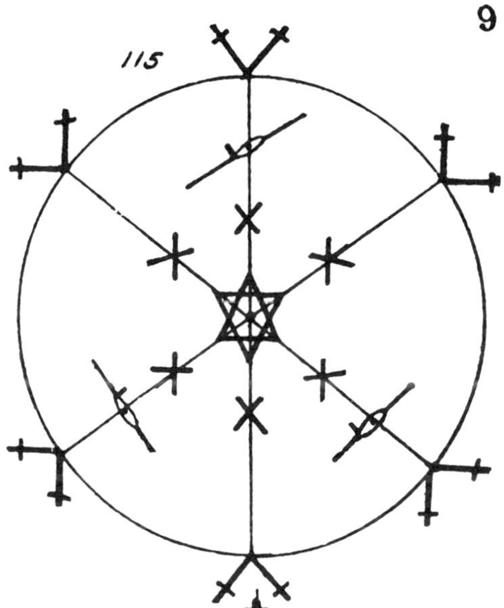

116

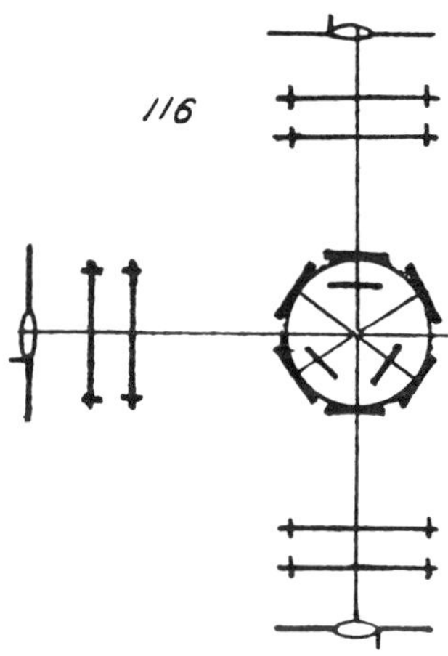

118

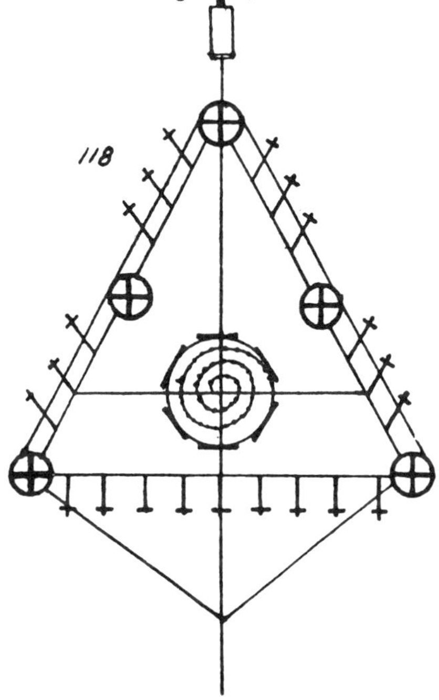

117

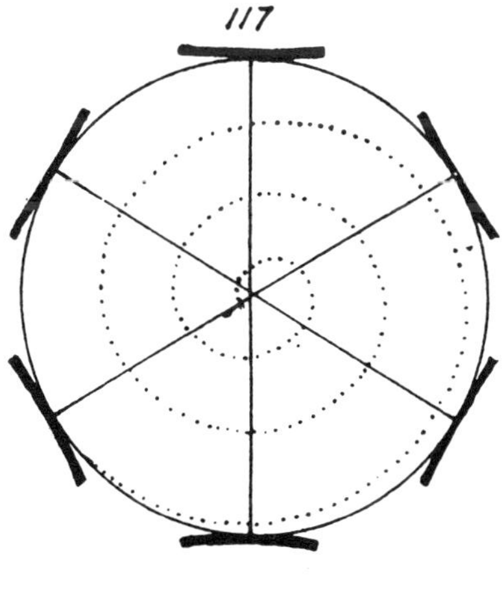

119

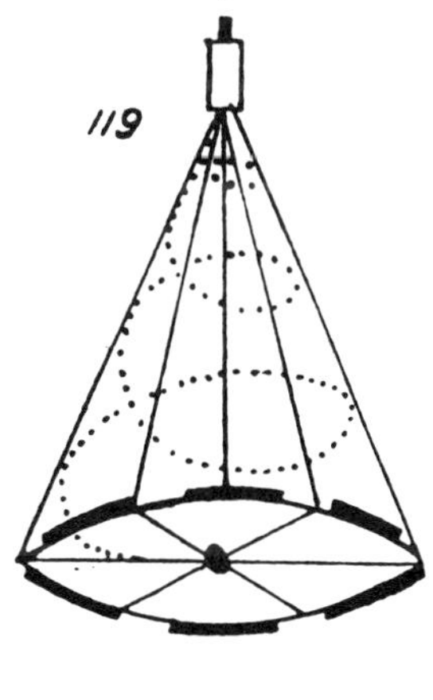

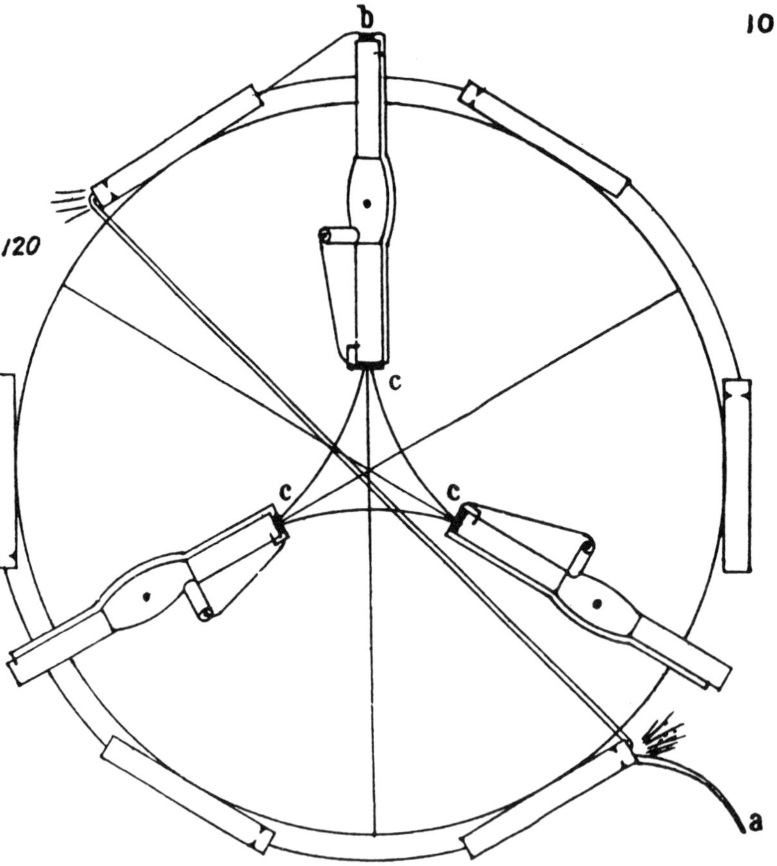

120

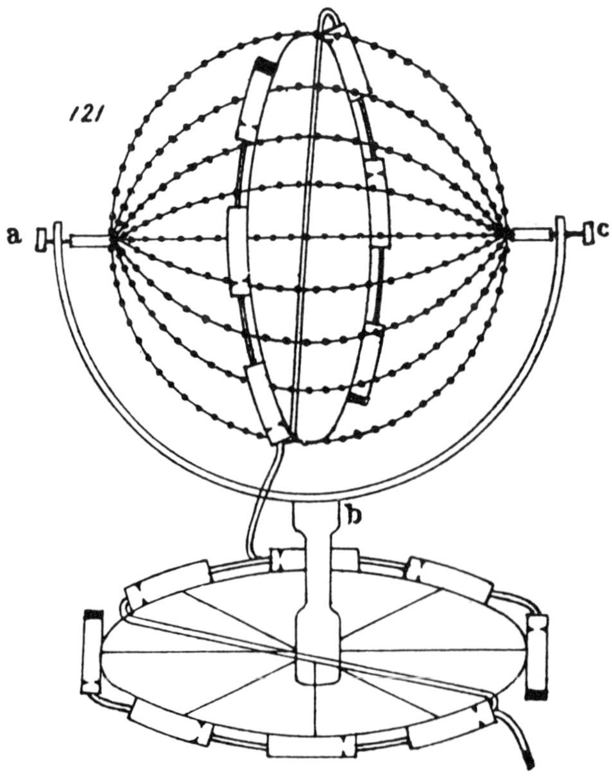

121

122

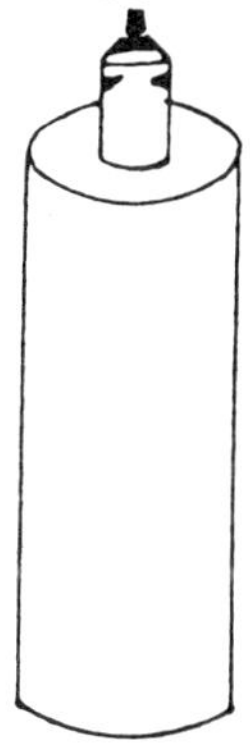

123

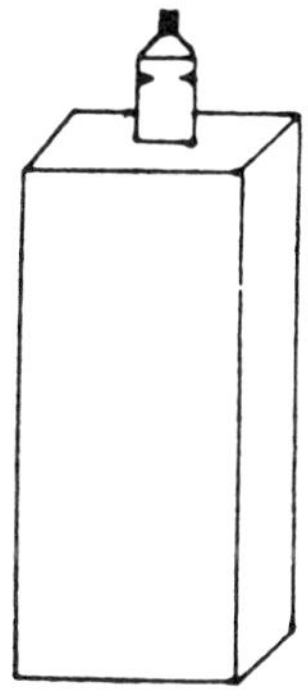

124

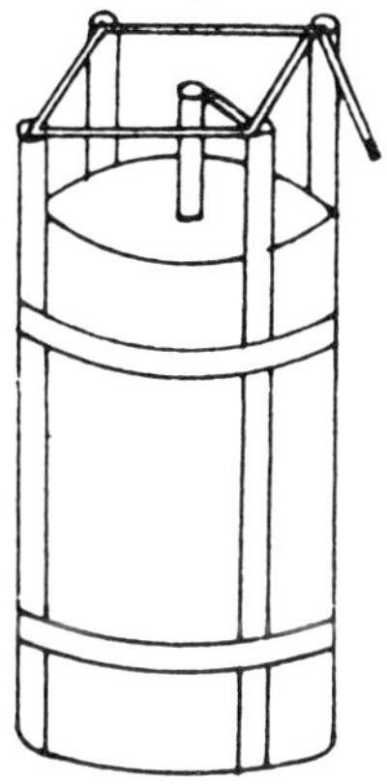

126

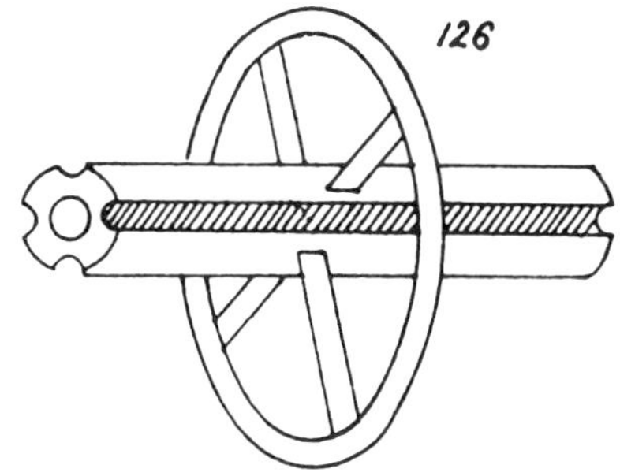

127

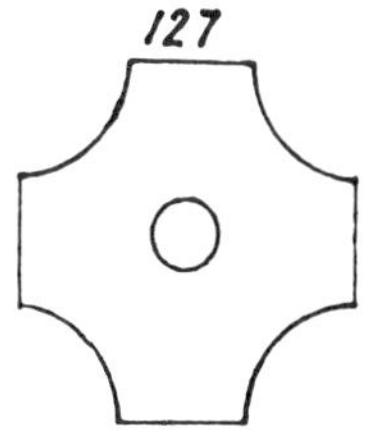

130

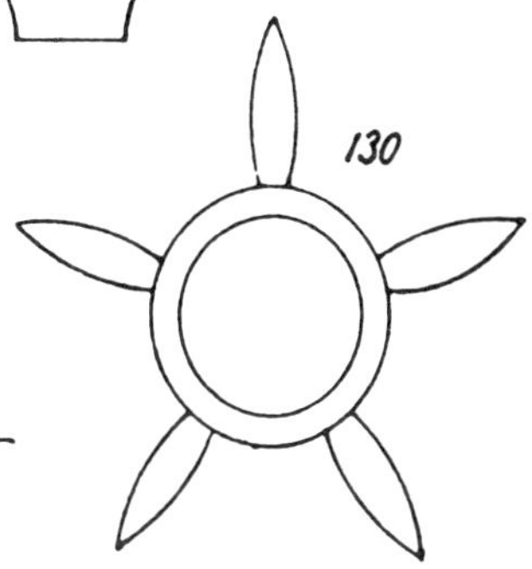

125

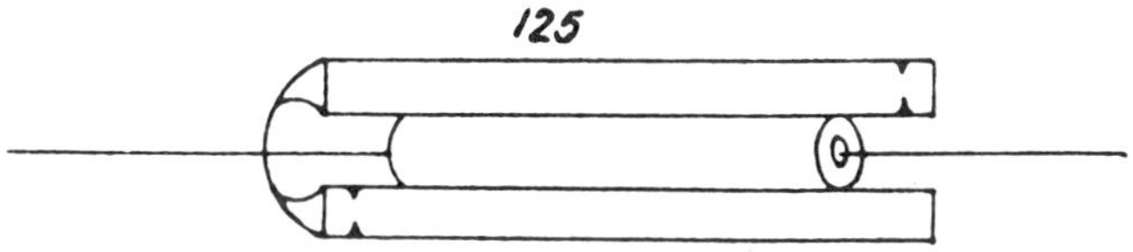

128

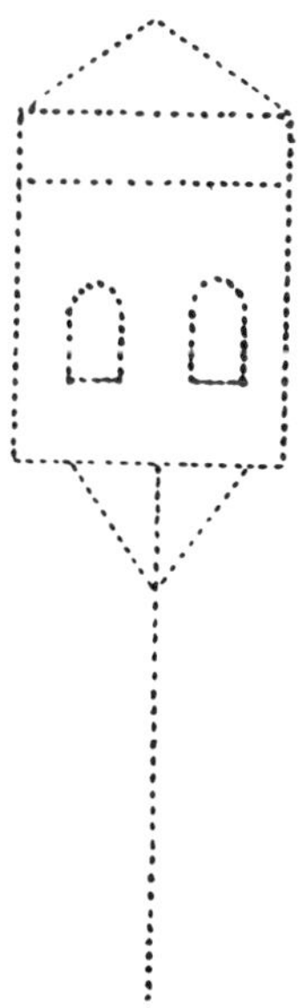

129

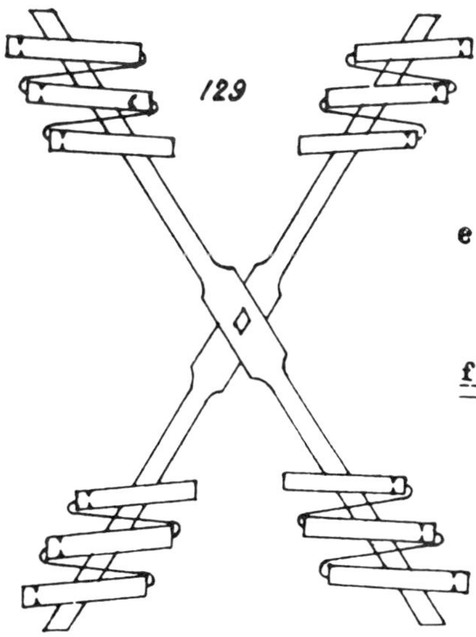

131

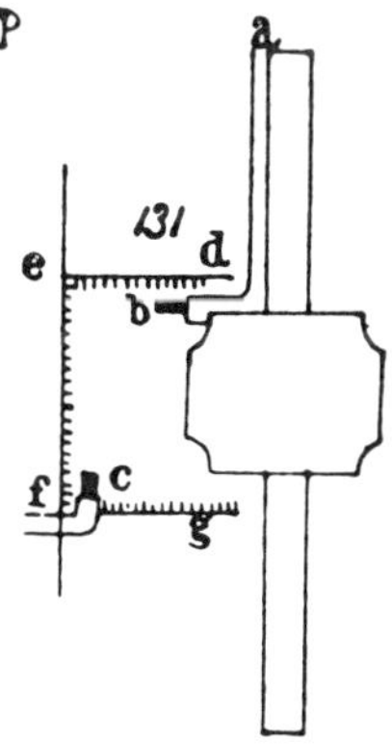

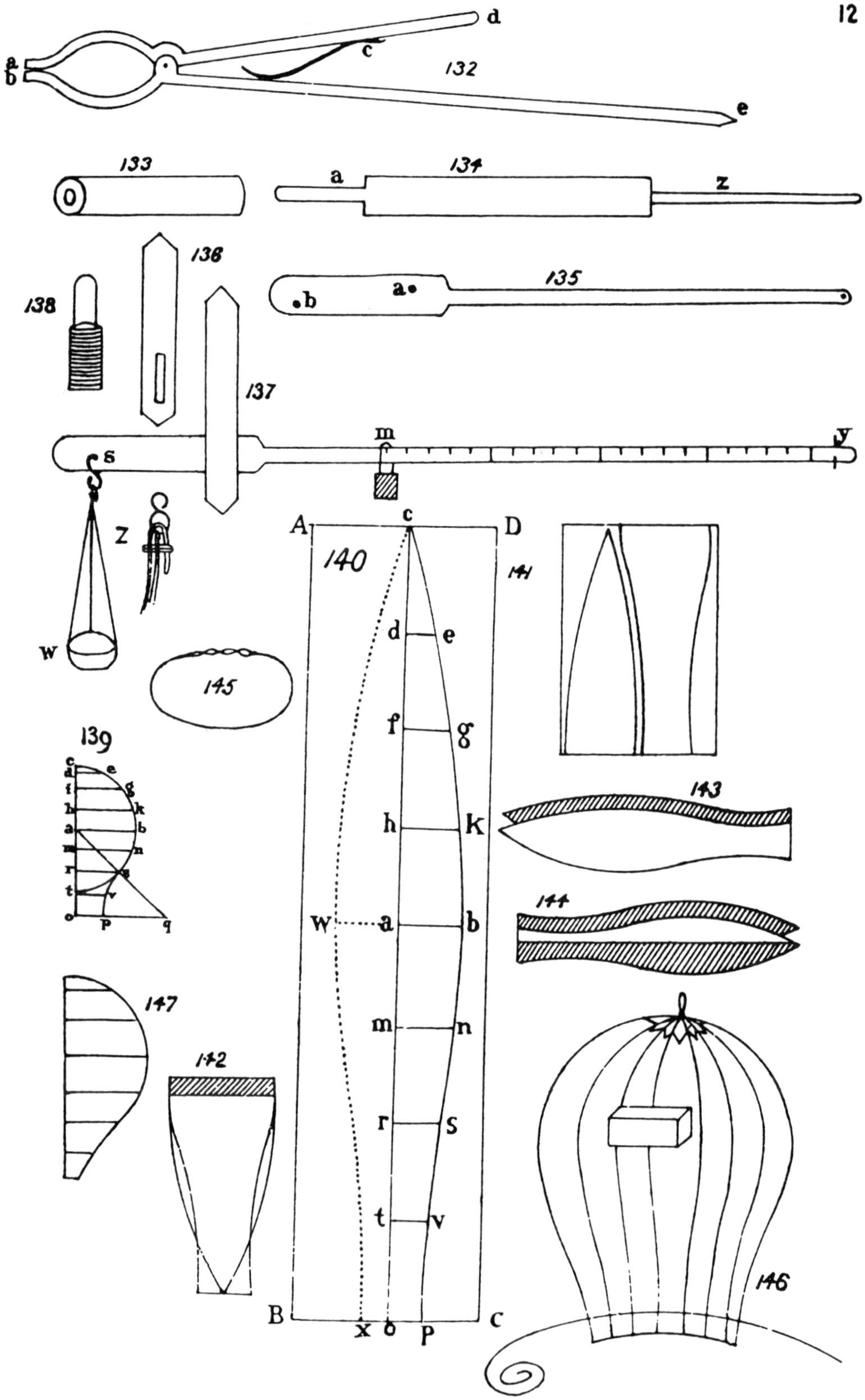
132
a
b
c
d
e
133
134
a
z
135
a
b
136
137
138
m
y
s
Z
w
140
A
c
D
B
C
x
o
p
W
d
e
f
g
h
K
a
b
m
n
r
S
t
v
141
143
144
145
139
147
142
146

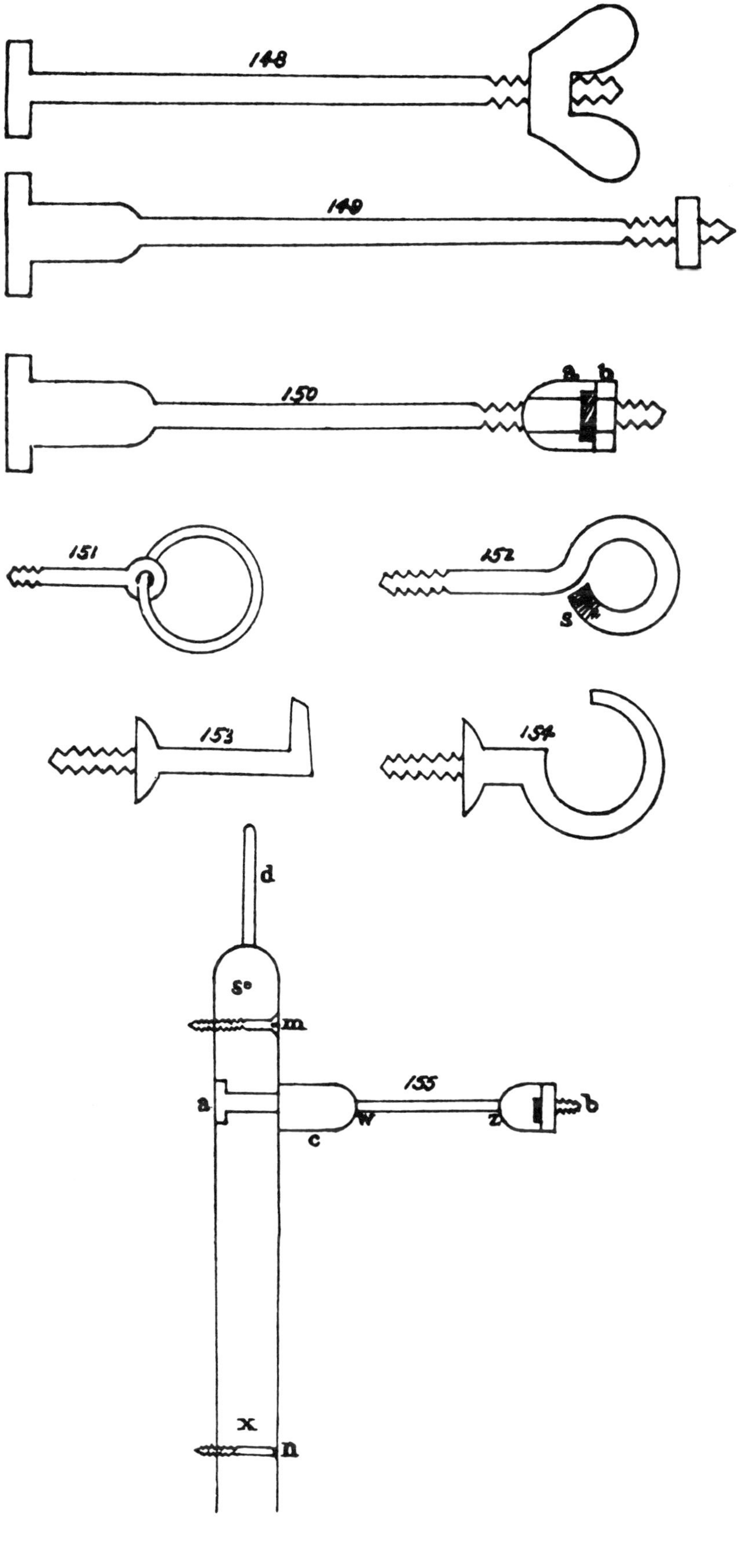
148
149
a
b
150
151
152
s
153
154
d
s°
m
155
a
w
z
b
c
x
n

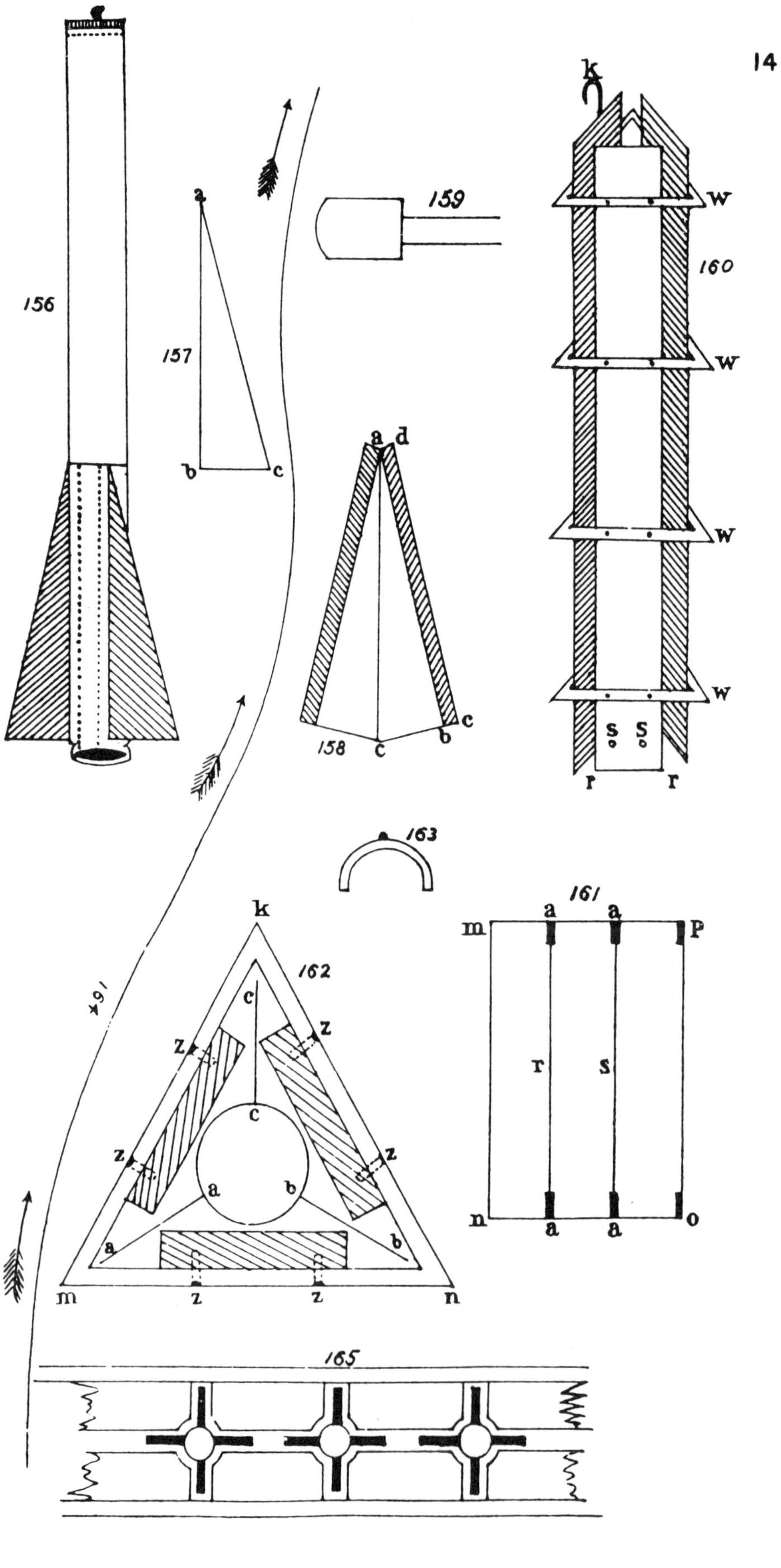
156
157
a
b c
158
a d
c b c
159
160
k
w
w
w
w
s s
r r
161
m a a p
r s
n a a o
162
k
c
c
z z
z z
a b
a b
m z z n
163
164
165

166

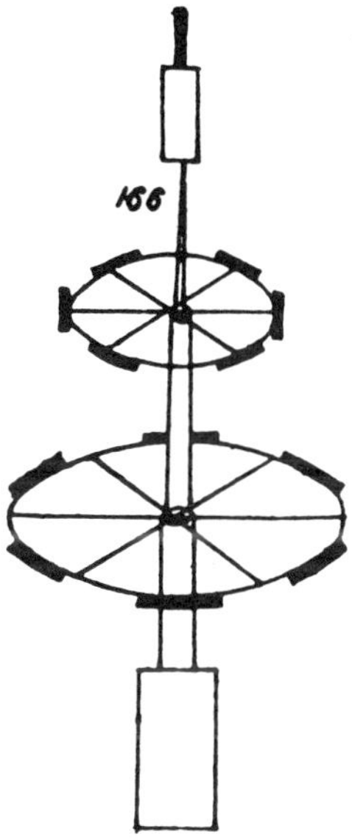

167

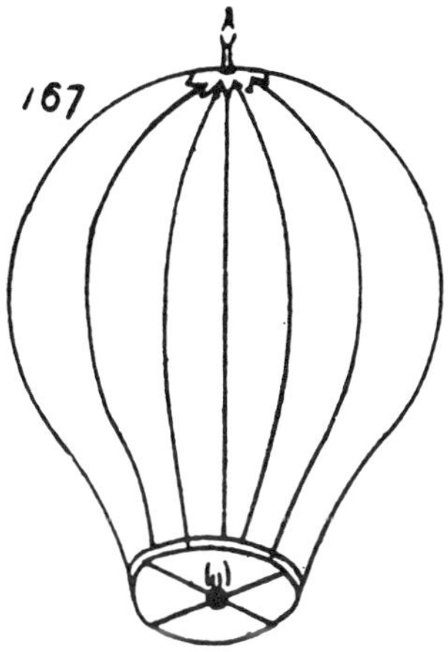

168

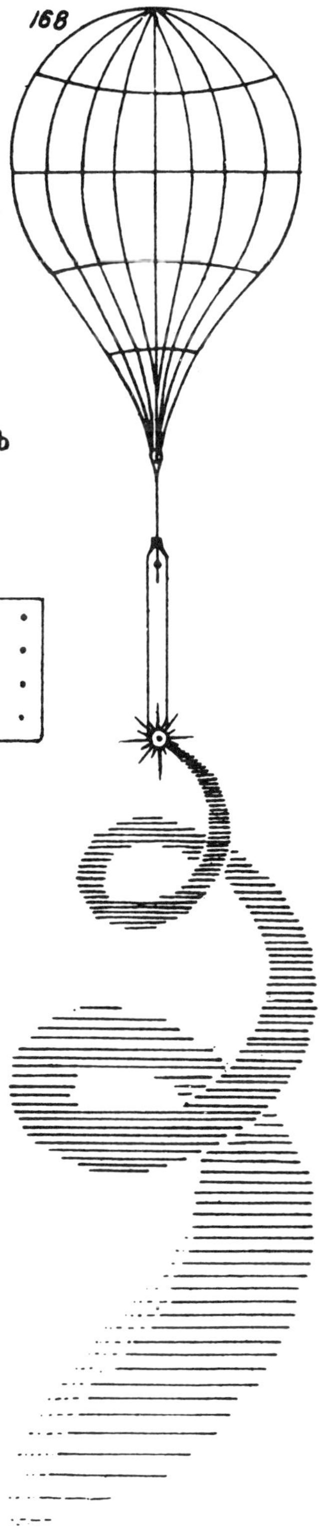

169

170

172

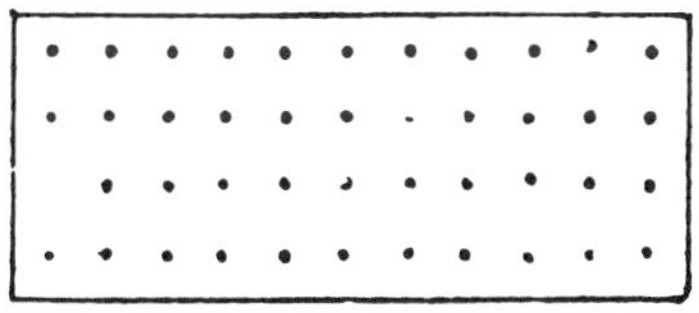

171

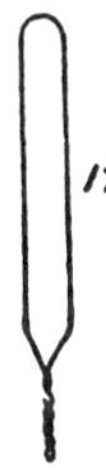

173

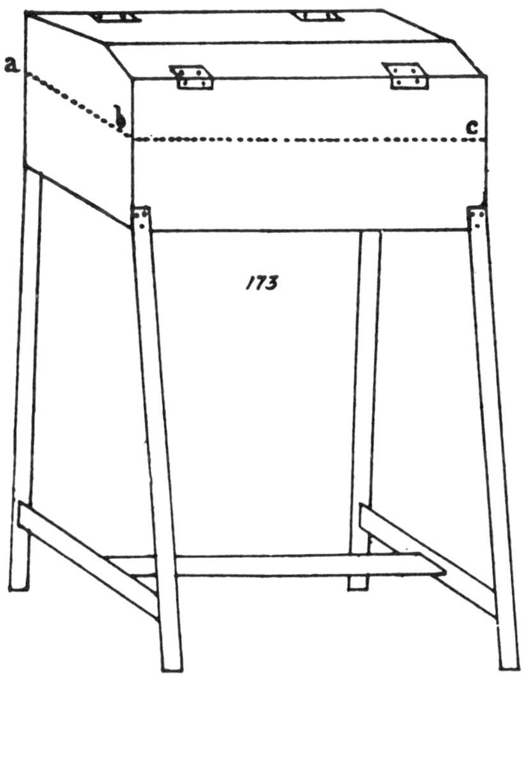

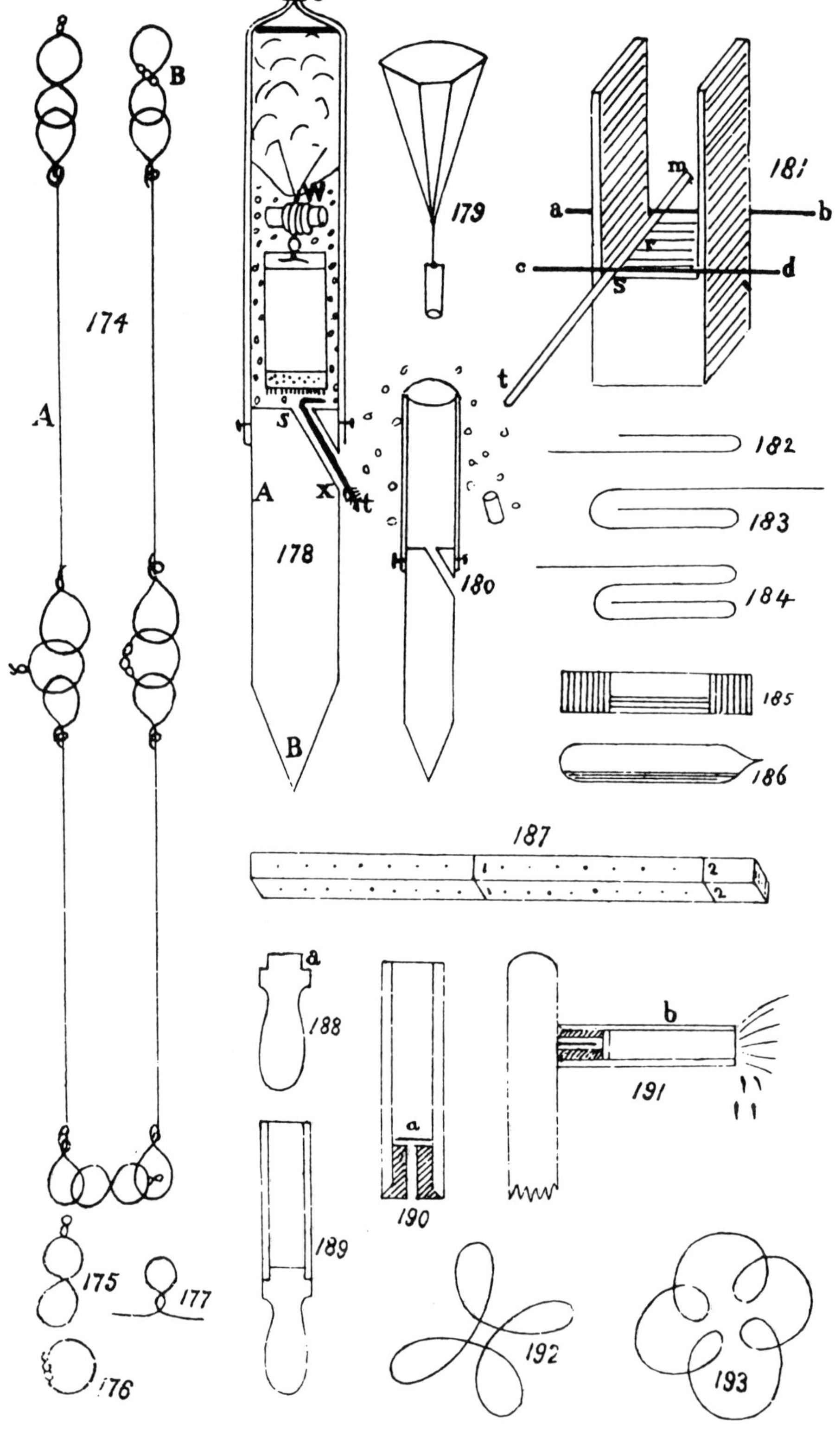
174
A
B
175
176
177
178
A
B
c
W
s
x
t
179
180
181
a
b
c
d
m
r
S
t
182
183
184
185
186
187
188
a
189
190
a
191
b
192
193

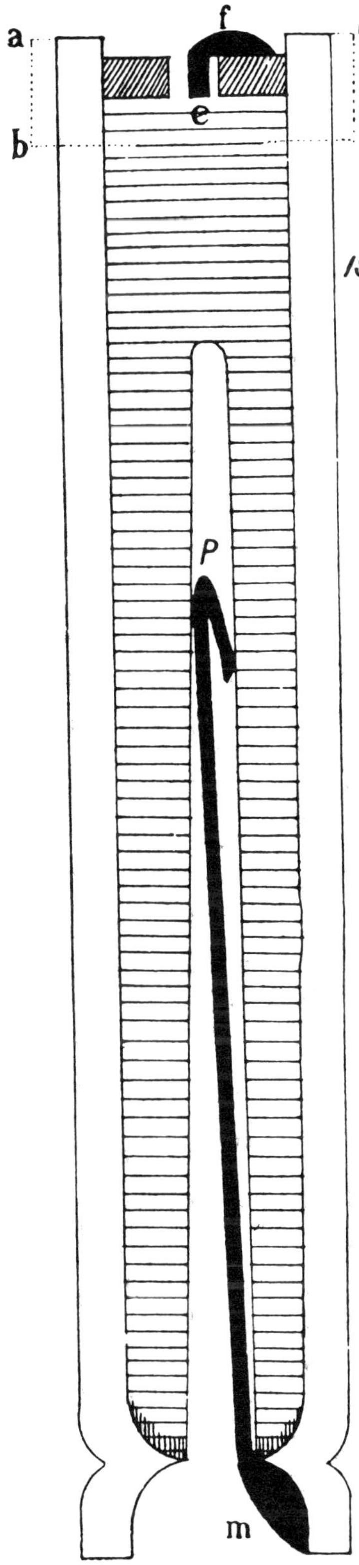

194

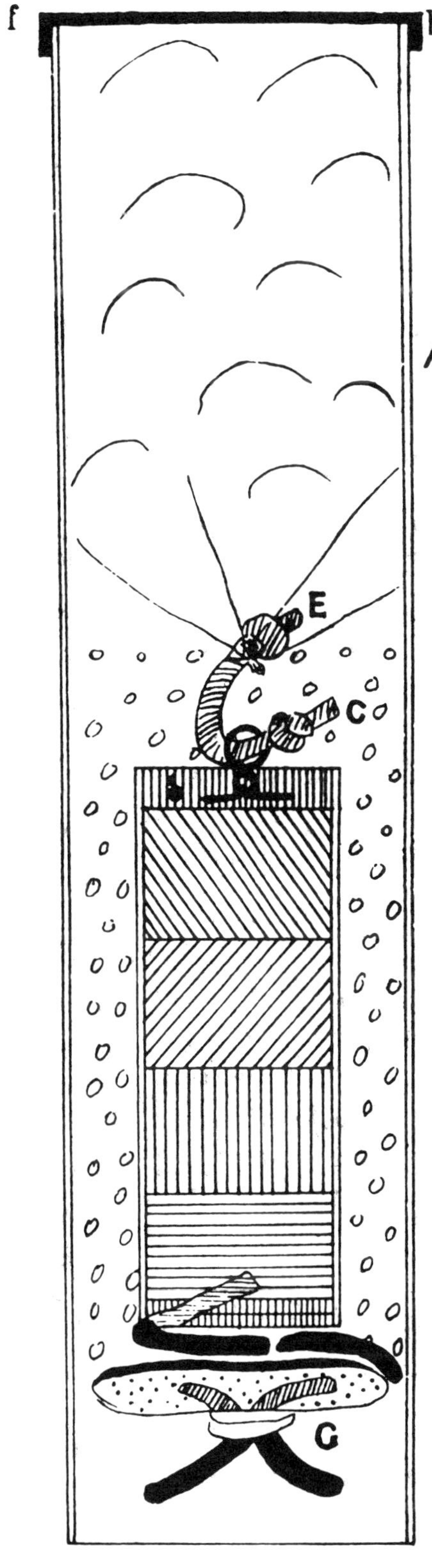

195

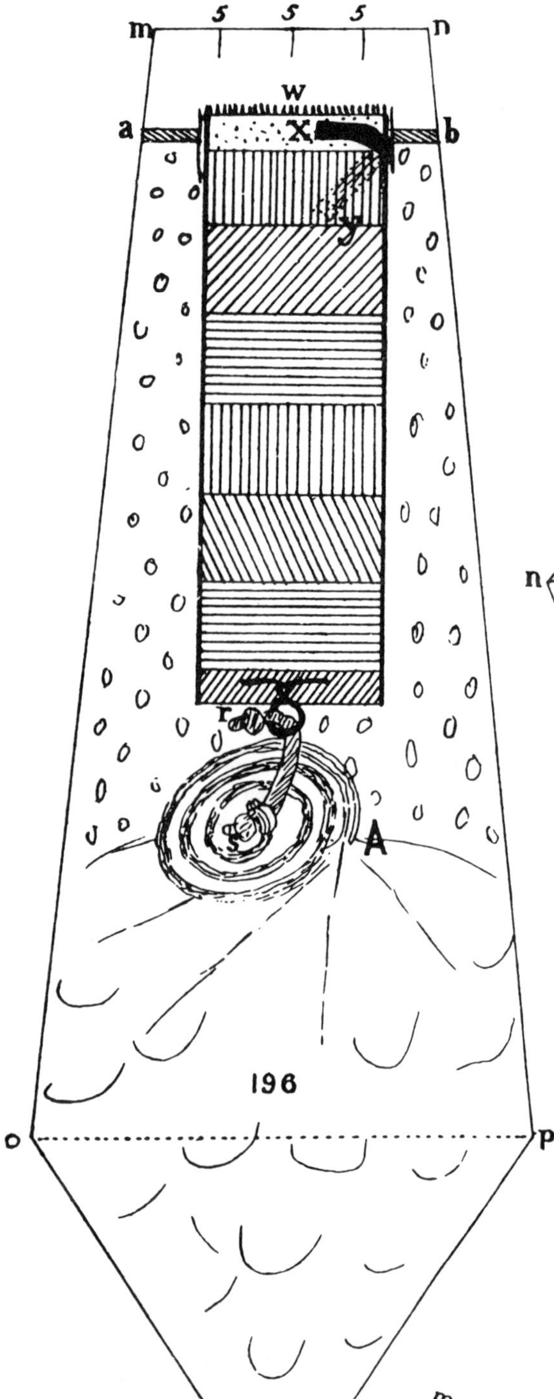

196

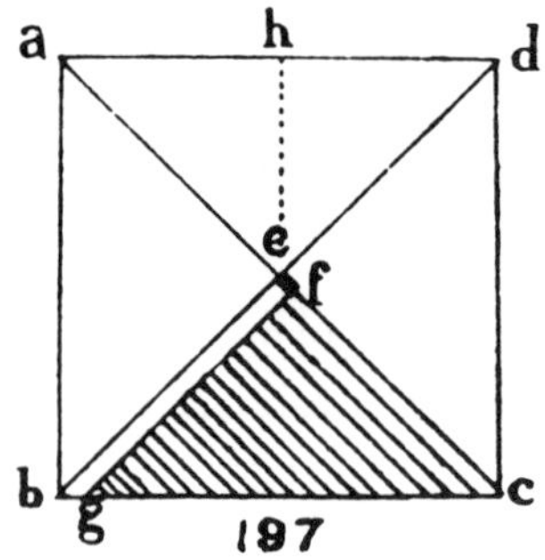

197

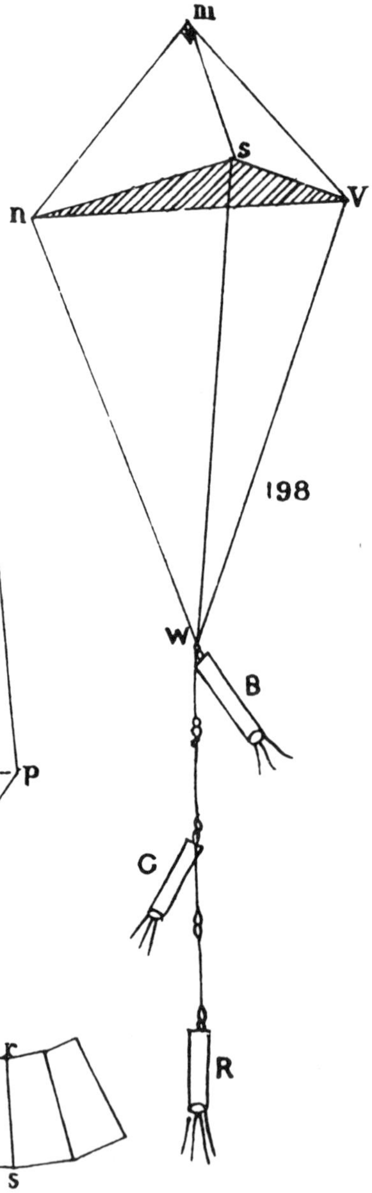

198

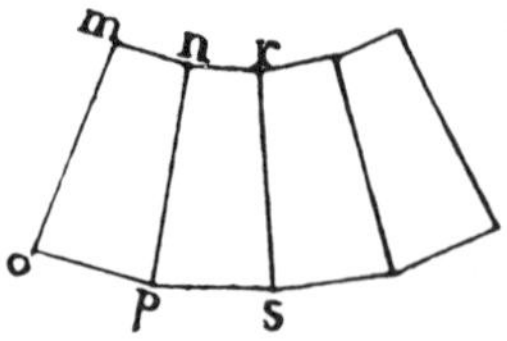

199

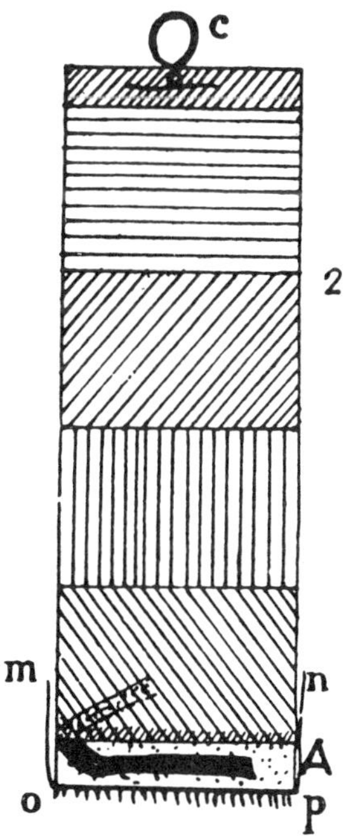

200

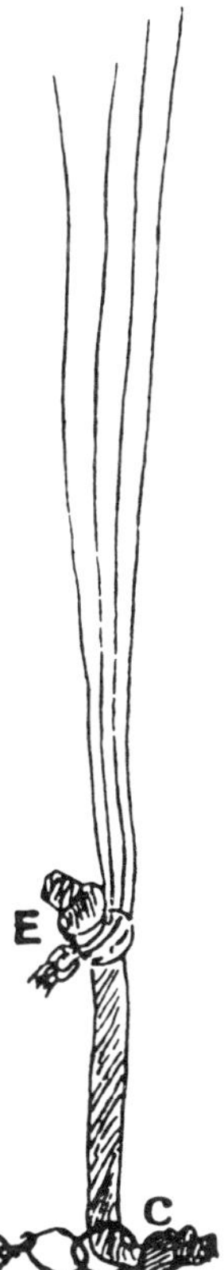

201

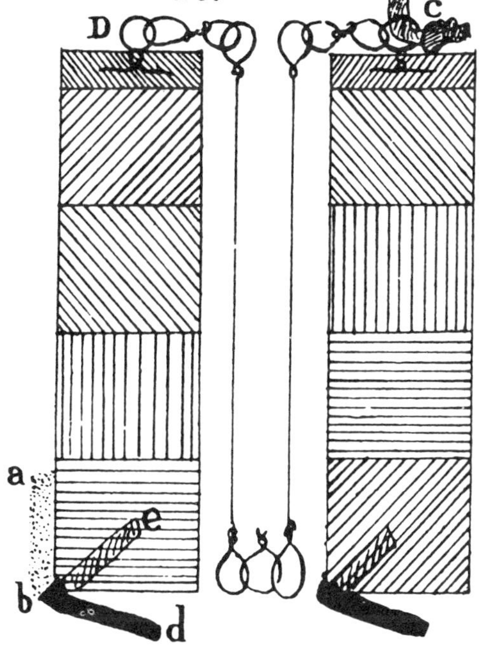

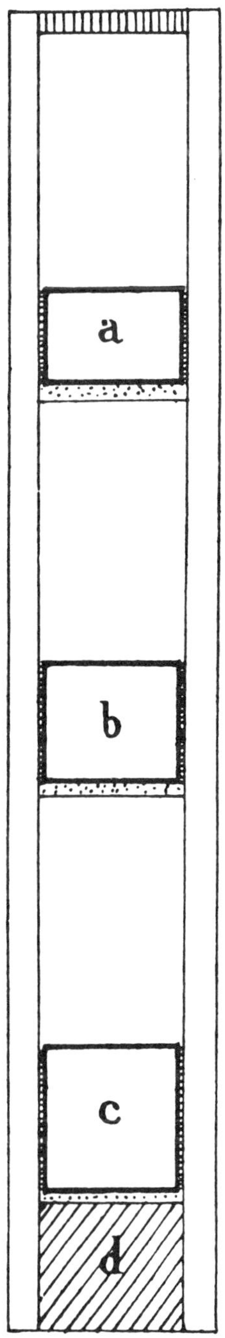

202

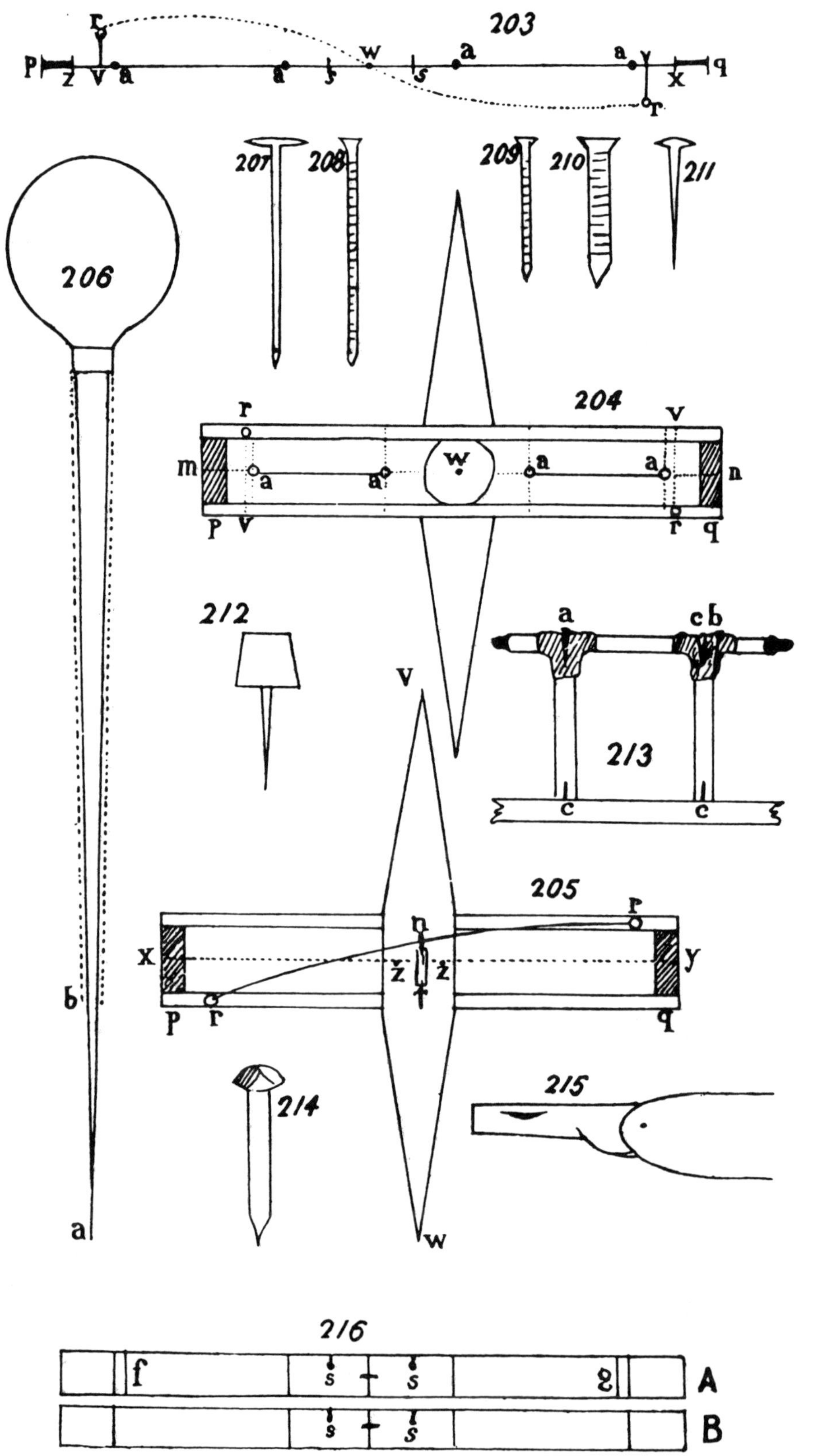
203
204
205
206
207
208
209
210
211
212
213
214
215
216
A
B

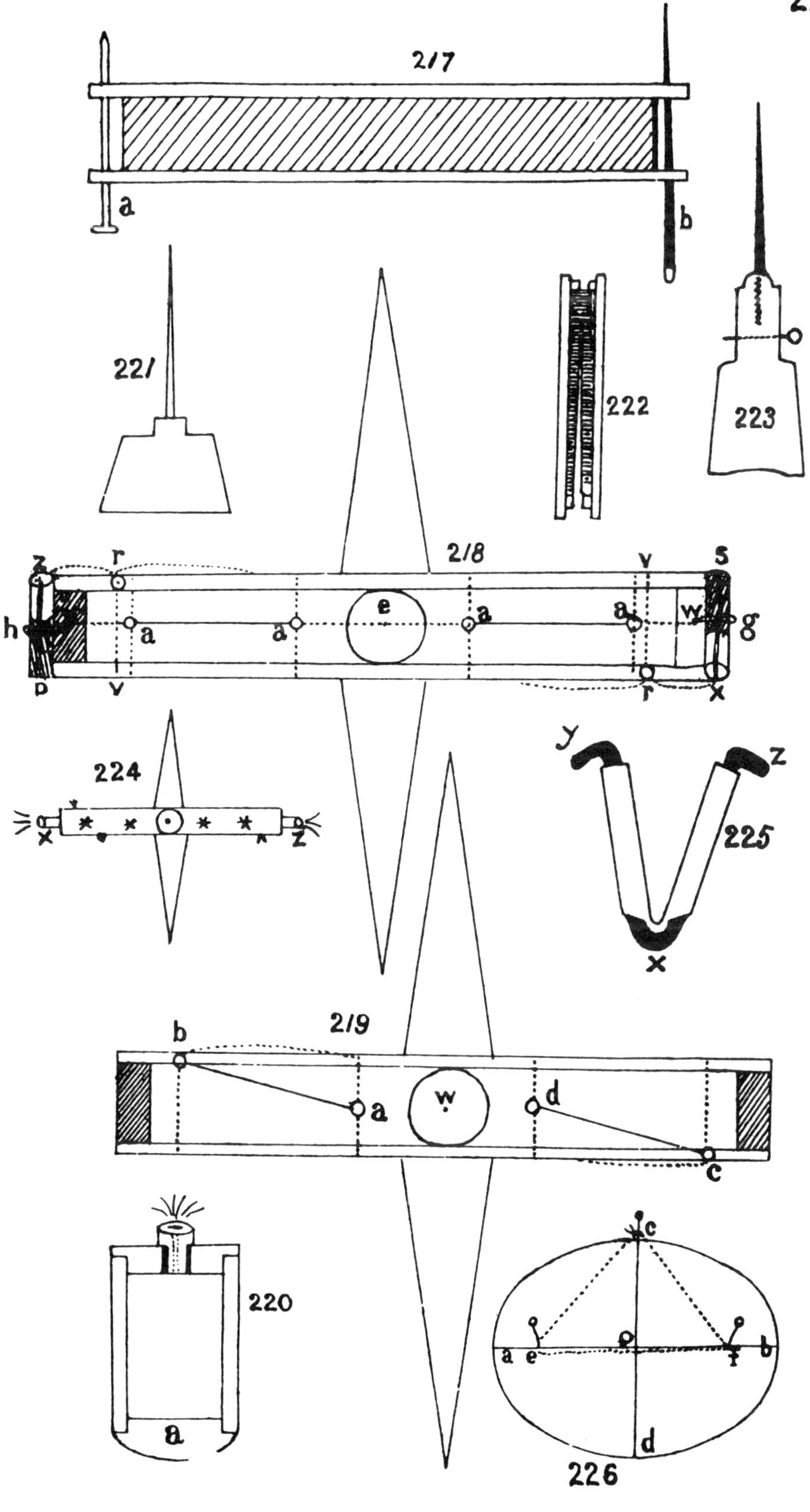
217
a
b
221
222
223
218
z r v s
h a a a a w g
p v r x
224
x z
225
y z
x
219
b
a w d
c
220
a
226
c
a e f b
d

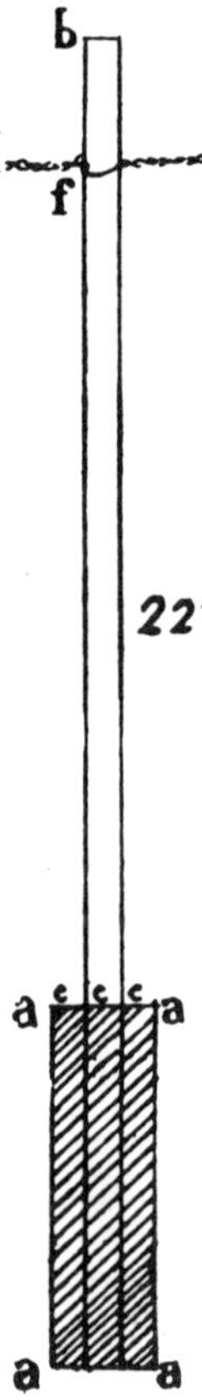

227

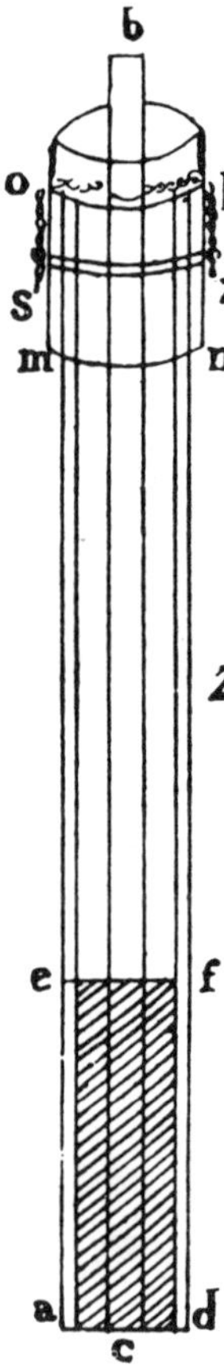

228

229

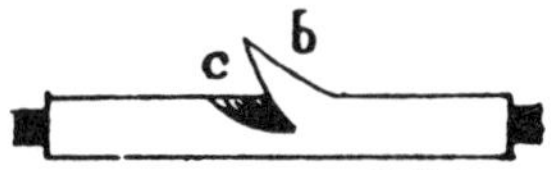

230

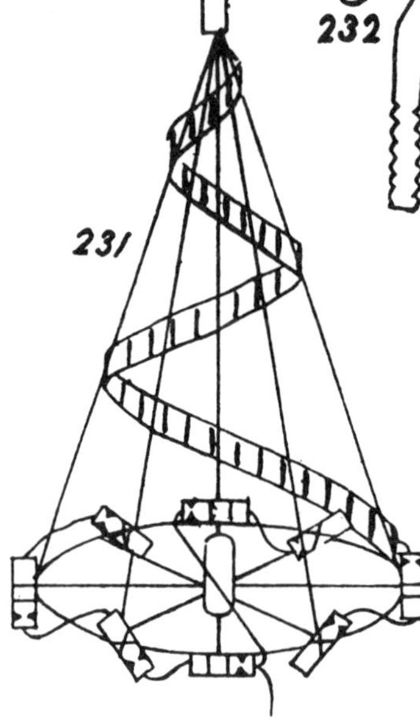

231

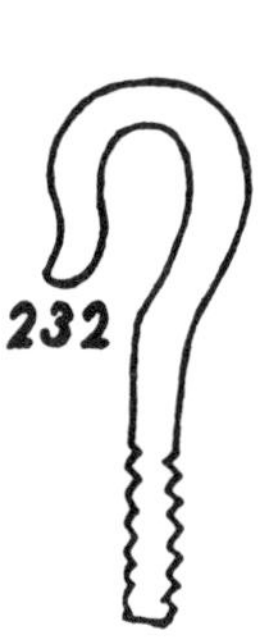

232

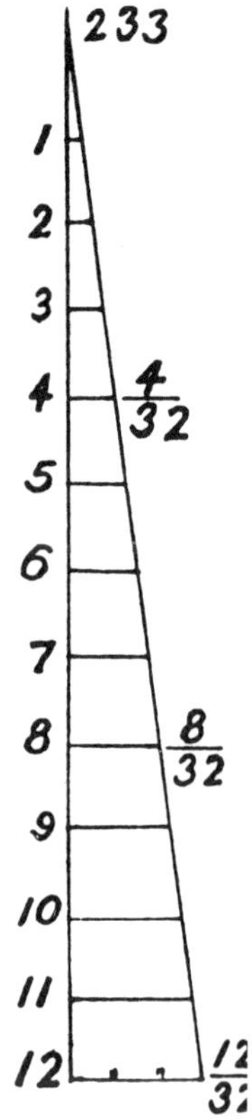

233

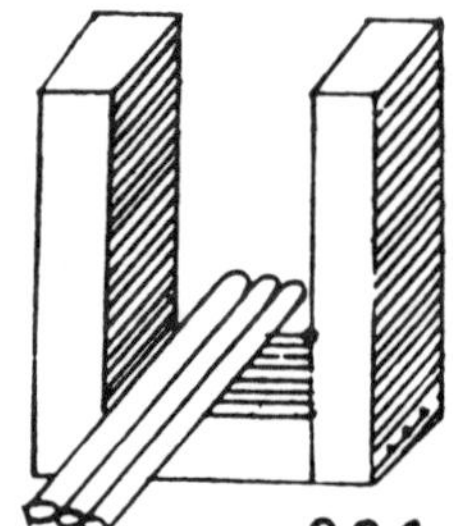

234

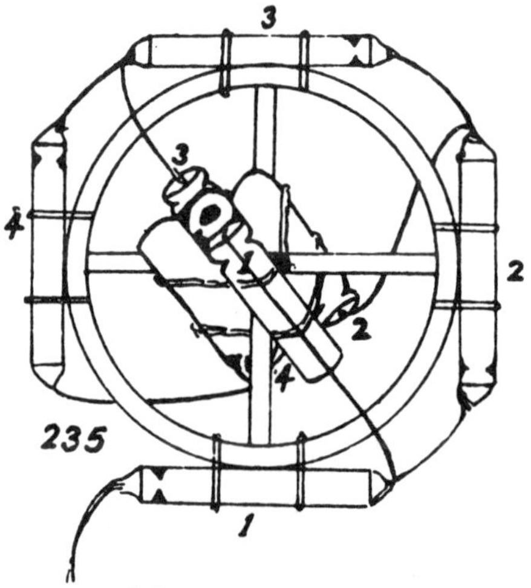

235

236

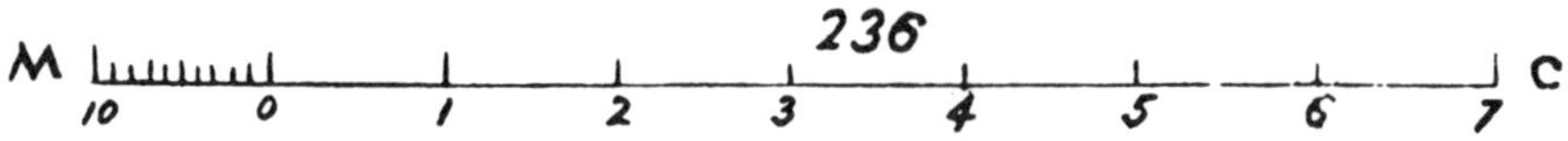

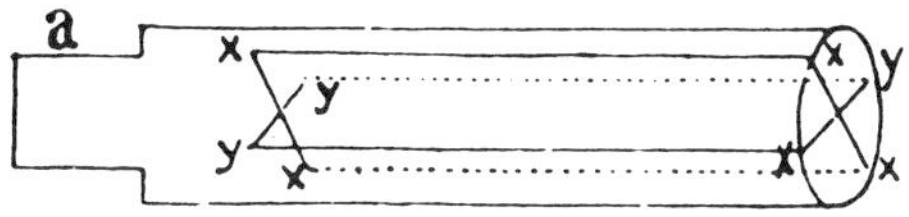

237

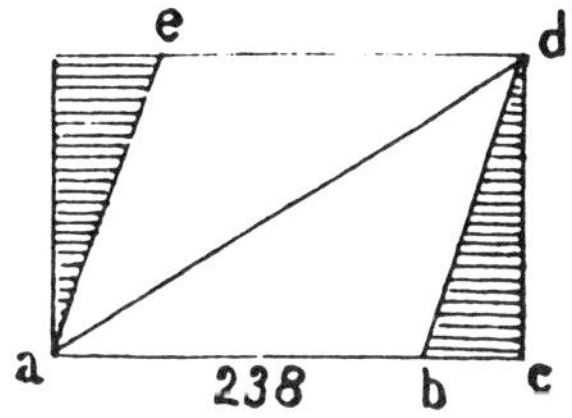

238

239

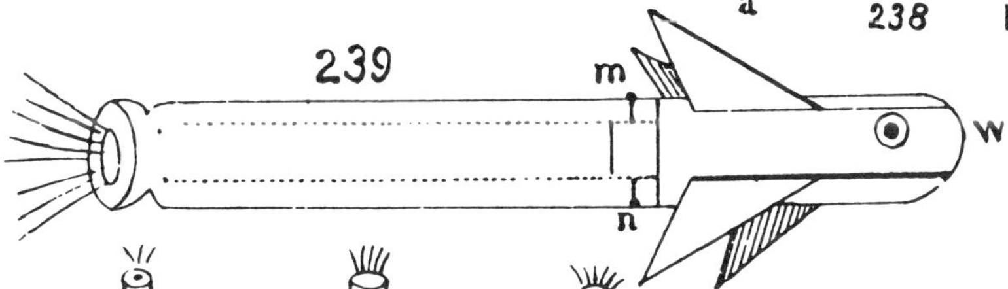

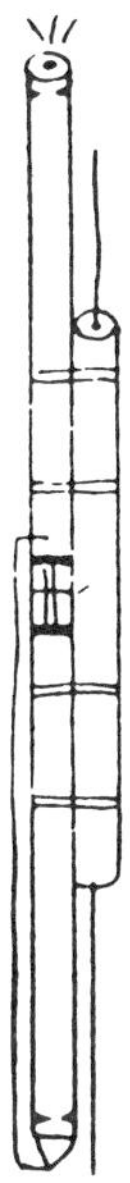

240

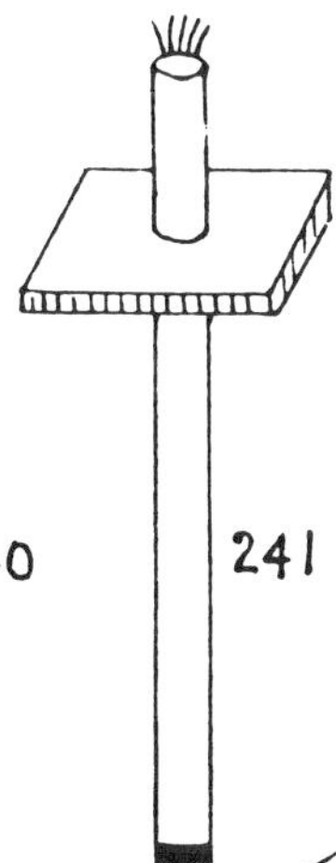

241

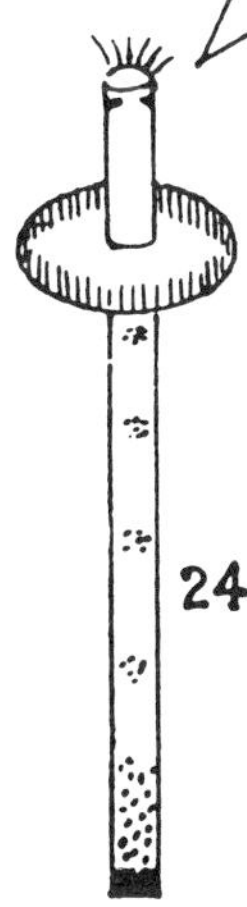

242

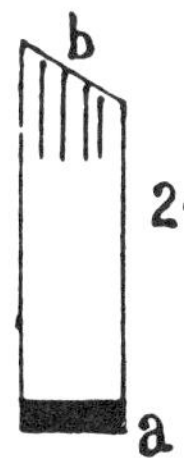

243

244

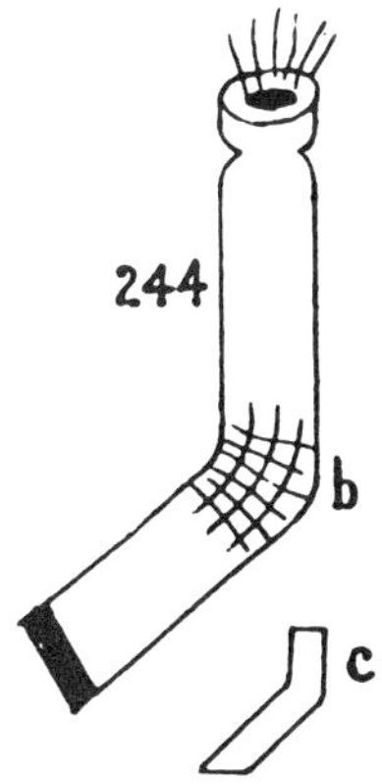

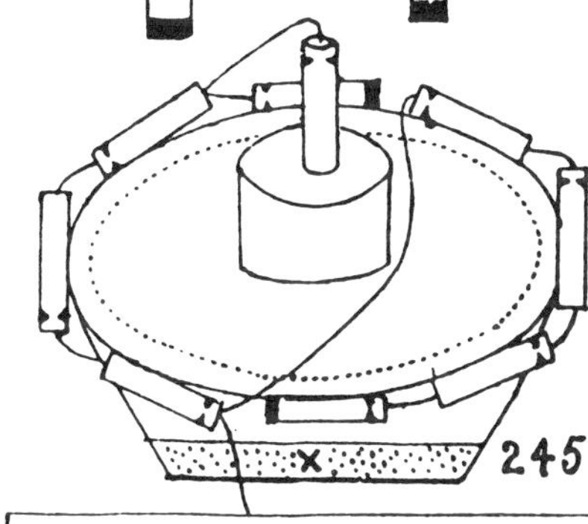

245

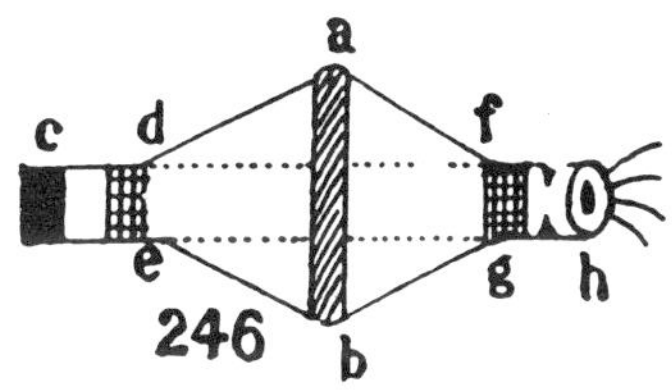

246

247

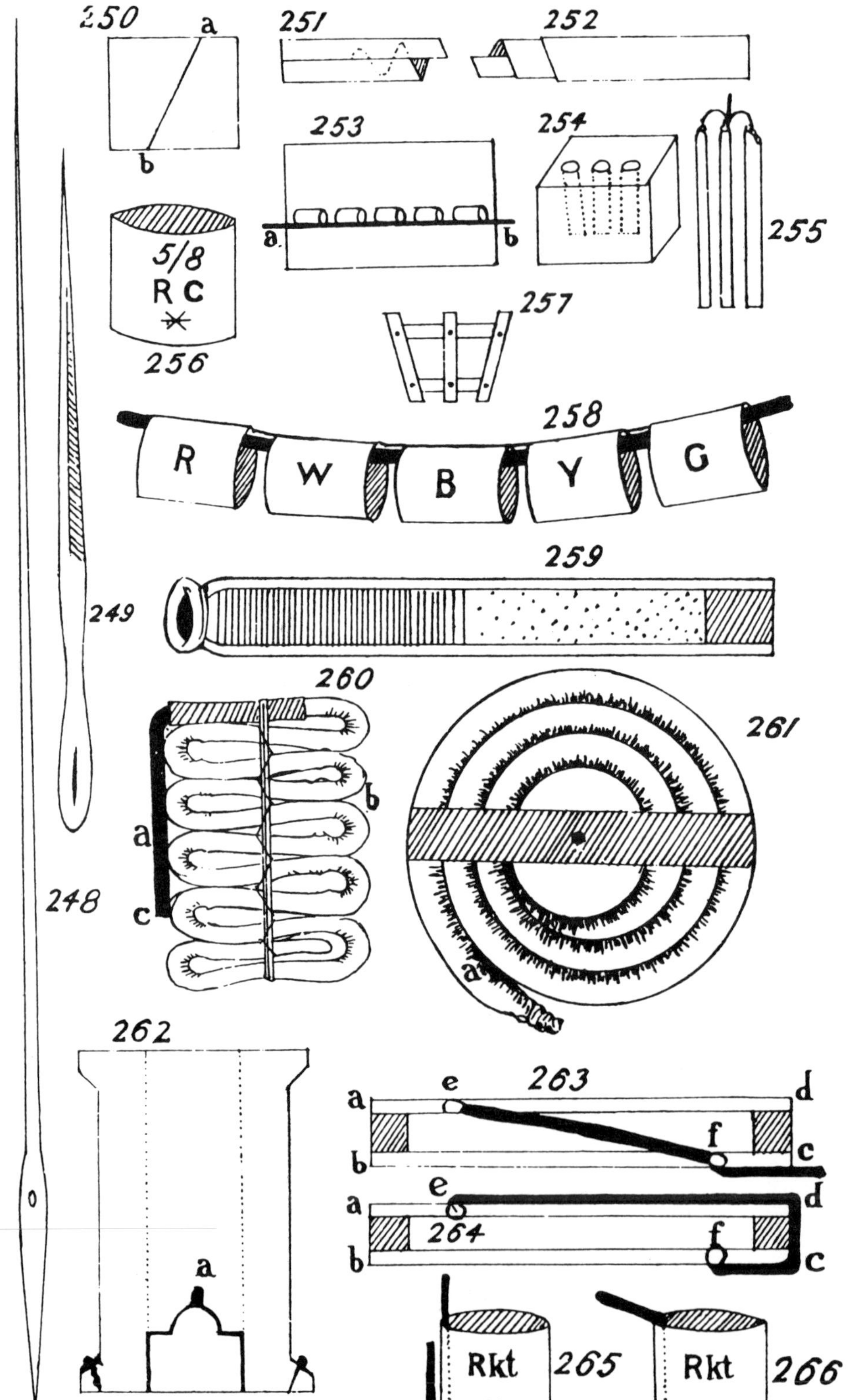
250
a
b
251
252
253
a
b
254
255
5/8
R C
256
257
258
R
W
B
Y
G
259
249
260
a
b
c
261
a
248
262
a
263
a
e
d
f
c
b
264
a
e
d
f
c
b
Rkt
265
Rkt
266